Implementing Six Sigma and Lean: A Practical Guide to Tools and Techniques

Ron Basu

Routledge
Taylor & Francis Group
LONDON AND NEW YORK

First published by Butterworth-Heinemann

Published 2014 by Routledge
2 Park Square, Milton Park, Abingdon, Oxon OX14 4RN
711 Third Avenue, New York, NY 10017

Routledge is an imprint of the Taylor & Francis Group, an informa business

First edition 2009

British Library Cataloguing in Publication Data
A catalogue record for this book is available from the British Library

ISBN: 978-1-8561-7520-3

Typeset by Charon Tec Ltd., A Macmillan Company.
(www.macmillansolutions.com)

To Indira and Mandira, my little angels

Contents

Preface

Background

Whilst passing through Miami airport en route Mexico City, I came across an article on 'Six Sigma' in *USA Today*, 21 July 1998. It read, *Today, depending on whom you listen to, Six Sigma is either a revolution slashing trillions of dollars from corporate inefficiency or it's the most maddening management fad yet devised to keep front-line workers too busy collecting data to do their jobs*. At that time I was coordinating a Global MRPII Programme between all manufacturing sites of GlaxoWellcome including the Xochimilco site in Mexico. The Global Manufacturing and Supply Division of GlaxoWellcome were considering a 'Lean Sigma' initiative which was meant to be a hybrid of Six Sigma and Lean Manufacturing. It struck me that the message in the *USA Today* reflected not just the doubts (or expectations) in the minds of my colleagues but perhaps those of quality practitioners worldwide.

These doubts or expectations addressed many questions. Isn't Six Sigma simply another fad or just a repackaged form of TQM? It appears to be successful in large organisations like Motorola and General Electric, but can small firm support such a programme? How can we apply Six Sigma methodology originated from manufacturing operations to the far larger market of the service sector? Like any good product, Six Sigma will have a finite life cycle – so what is next? Surely one big question is how can we sustain the benefits in the longer term? It is good to be 'lean', but isn't it better to be 'fit' to stay agile? The idea of writing a book to address these issues was mentally conceived at Miami airport and the concept of FIT SIGMA™ was born. It was named *Quality Beyond Six Sigma* and the book was published by Butterworth Heinemann in January 2003.

During the preparation of *Quality Beyond Six Sigma* and particularly after its publication it struck me that the book fell in the same trap like other books on quality genre. The emphasis has been towards the strategy and culture of change management. The fundamental building blocks of hands-on tools and techniques were missing. The wisdom of a people focused holistic approach to operational excellence cannot be faulted. It is vital to follow a proven approach to put all the components of the quality improvement puzzle together properly. It is equally important to explain the tools and techniques in detail to all stakeholders of the programme so that these components can be applied in the most effective way. One can argue that the motivation to effectively pursue quality improvement has been impaired by not fully understanding the basic tools and techniques. Furthermore there is no available publication which is dealing in-depth with these useful tools and techniques under one cover.

In the experience of the author, both in industry and management courses, the details of various tools and techniques had to be acquired from different books, publications and training manuals. It is time now to produce a comprehensive, user-friendly and hands-on book which could be a single source reference of tools and techniques for all practitioners and students of operational excellence. With these thoughts in mind I wrote *Implementing Quality*, which was published by Thomson Learning in 2004. The demand for this book, comprising tools and techniques and implementation methodology for quality programmes, such as Six Sigma, Lean Sigma and FIT SIGMA™, continued to grow. However during 2007 Thomson Learning, following a takeover and organisational restructuring, changed their strategy of publications and Elsevier expressed interest in publishing an updated version. So this book, *Implementing Six Sigma and Lean* was born.

Tools and techniques

I firmly believe that regardless of the quality programme that an organisation may choose to adopt, a selection of tools and techniques will be essential to progress the initiative. You should have a complete toolbox at hand. Just as a good golfer will not compete in a championship with just an iron, it is important that you should acquire a full bag of irons and clubs. The reader should find a comprehensive range of tools and techniques and useful management models in this book.

The terms 'tools' and 'techniques', which are frequently and interchangeably used in books and practices, are consciously differentiated here. As described in the book. 'tools' are related to improving quality as are the tools used in constructing a building. A single 'tool' may be described as a device which has a clear role and defined application. A 'technique', on the other hand, may be viewed as a collection of tools. There is also a need for a greater intellectual thought process and more skill, knowledge, understanding and training in order to use them effectively. For example Control Charts is a tool of statistics. The way this tool is used along with other tools (e.g. Process Capability Measurement) is a 'technique' of Statistical Process Control (SPC).

It goes without saying that by borrowing the racket of Pete Sampras does not make one to win a Wimbledon title. Without appropriate training and application, the tools and techniques cannot be effective. It reminds me of a story that I read in 'Gems of Educational Wit and Humour'. A Baby Tiger said to Mummy Tiger, Here comes a hunter, and he has five rifles, three special sighting telescopes and devices to allow him to see in the dark! What are we going to do'? 'Hush!' replied Mummy Tiger and taught her cub how to sneak up from behind and pounce. The hunter was never heard of again.

A simple moral from this story is that tools and techniques may be fine, but these will never be a substitute for a good basic education.

About this book

The major new features of this book as compared to *Implementing Quality* include:

- Review of new trends such as Digital Six Sigma
- Force Field Diagram in Chapter 7
- TRIZ in Chapter 6
- Lean Thinking in Chapter 10
- Introduction to basic statistics in Appendix 2.

This book is about a practical guide to tools and techniques which are necessary in the implementation of all kinds of quality programmes at whatever level or under any banner these may be pursued, such as Continuous Improvement, TQM, BPR, Six Sigma, Lean Enterprise, Lean Sigma, FIT SIGMA™, to name a few.

The approach of the text focuses on the clear definition, description and application of each tool or technique underpinned by numerical or practical examples. Selective worked-out examples in the text would help readers with some grounding in real metrics which they can subsequently apply in their own organisation.

The text covers an implementation programme to demonstrate how to put the pieces of tools and techniques together. A differentiating feature of the approach is the end-of-chapter questions and exercises to help both directed and self-assessed learning.

The case studies at the end and the text address both services and manufacturing and provide a broad and practical perspective in four parts:

- *Part 1: The Foundations of Quality.* Deals with some of the fundamentals and the perception of quality which varies significantly depending on the type of industry, economy and culture.
- *Part 2: Tools.* Provides the definition, application, examples and training requirements of tools. It contains five chapters structured in the sequence of probable tools used in DMAIC.
- *Part 3: Techniques.* Gives complete coverage of advanced quantitative techniques and team based qualitative techniques supported by examples, training requirements, benefits and pitfalls.
- *Part 4: Implementation.* This part provides a step-by-step approach of making it happen in all types of organisation. Case studies are included to offer a practical insight of their applications.

Who should use this book

This book is aimed at a broad cross-section of readership including:

- Functional managers, participants and practitioners in TQM, Six Sigma and Operational Excellence will find this book will provide them with a comprehensive insight into tools and techniques of continuous improvement in one package. A step-by-step guide is included for the application of the appropriate tools to their improvement processes. This book could be used as an essential handbook for all employees in a Six Sigma programme.
- Senior Executives, both in the manufacturing and service industries (regardless of function) will find that this book will give them a better understanding of basic tools and techniques and help them to support a quality improvement initiative and sustain a strong competitive position.
- Professional management and training consultants will find the comprehensive approach of tools and techniques as an essential handbook for Six Sigma related assignments and seminars.
- Management schools and academies and research associations will find this book valuable to fill the visible gap in basics of operational excellence. This text will provide support to both undergraduate and post graduate courses containing quality and operational excellence and as a main textbook for the Quality Elective for MBA students.

The readership will be global and particularly cover North America, the UK, Continental Europe, Australia and Asia Pacific countries.

How to use this book

Application

The book allows for maximum flexibility for readers and users to apply it depending on their requirements and interests. The application areas of the book include the following.

Implementing Quality Six Sigma, FIT SIGMA™ and Lean

The organisations, whether services or manufacturing, private or public sectors, large, medium or small, should particularly benefit from the success factors in Chapter 3, the methodology of implementation in Chapter 11. The programme members and task groups should acquire a copy of the book and gain a common understanding of tools and techniques described in Parts 2 and 3. Gems of Educational Wit and Humour (Mamchak and Mamchak (1999)).

Developing a training programme

The definitions, basic steps, examples and training requirements for each tools and techniques which are described in detail are the building blocks of the training modules of a Six Sigma or related training programme. Specific guidelines have been included to select the tools and techniques for specific levels of training, such as Black Belts, Green Belts and awareness training.

University and college courses

The book can be as a textbook or a reference book for advanced programmes in Managing Quality in Universities and Business Schools. Questions and exercises after each section should help students the chance to practice and assess their level of understanding achieved from the relevant chapters. The tutors will have the opportunity of applying these questions and exercises as part of their lecture materials and course contents.

Handbook of tools and techniques

The book contains a comprehensive compilation of tools and techniques in Parts 2 and 3 and management models in the Appendix 1. There is also a Glossary of relevant technical and business terms. This is a rare collection of tools and techniques under one cover.

Enhancing knowledge

The book contains both the strategic approach of implementing quality and the detail coverage of tools and technique which underpin the programme. The reader, whether a CEO, employee or a student, should find the book as a self-help of enhancing the knowledge and understanding of quality and operational excellence.

Key features

Foundations of quality

At the beginning of the book the reader finds a good understanding of what is meant by quality and its dimensions, the key messages from the experts or 'gurus' of quality movement and the scope of tools and techniques. This provides a solid starting point.

Summary

At the end of each chapter you will find a summary of key elements covered. This should provide a sound basis for general revision.

Basic steps and worked-out examples

All tools and techniques described in the book contain definition, application, basic steps and worked-out examples. Each demonstrates a hands-on concept providing the reader with a step-by-step clarification and understanding.

Training requirements

The specific requirements of training depending on its application are included for each tool or technique. This also provides an indication of the educational background or skill level required for a user for the success of its application.

Benefits and pitfalls

For all techniques both the benefits and pitfalls have been clearly stated so that a user can obtain a balanced view of the technique without being unduly influenced by its popularity, fashion or fad.

Final thoughts

I have included my personal tips for each tool or technique for the reader to consider. This has been covered under 'final thoughts'.

Questions and exercises

As mentioned earlier questions and exercises after each section offer both the students and tutors an opportunity to practice the appropriateness of concepts in academic applications.

Case studies

The case studies have been included to encourage the readers to respond in practical situations. These are concisely written and provide a good learning resource for tutorials.

Further readings

General recommendations for further reading are included in each chapter in References for specialised books and many journal papers.

Management models

The Appendix contains a collection of management models which should be of special interests for Senior Managers in industries and business school students. This is also useful in the development of specific business strategies.

Glossary

A comprehensive Glossary of relevant terms has been provided at the end of the book. This provides a ready reckoner for common terminology and phrases experienced in managing quality.

I have made effort to provide you both simple and complex concepts easy to understand and with enough common sense you can apply them readily to make *Implementing Six Sigma and Lean* a reality in your organisations and programmes. Mahatma Gandhi once said, 'Be the change you want to be in the world'. I hope this book will in some way help to bring that change.

Ron Basu
Gerrards Cross, England
January 2008

Acknowledgement

I acknowledge the help and support from my colleagues and students at Henley Management College, especially Prof. Peter Race, and Essex University in England and ESC Lille in France. Always it has been my pleasure to work with Dr Nevan Wright, my co-author of a few other books. I gratefully acknowledge Nevan's contribution particularly for Chapters 2 and 11.

My sincere thanks go to the staff of my publishers especially to Maggie Smith of Elsevier and Jennifer Pegg of Cengage Learning for getting this project off the ground.

Finally the project could not be completed without the encouragement of my family, especially my wife Moira and daughter Bonnie.

Ron Basu

About the author

Ron Basu is Director of Performance Excellence Limited and also a Visiting Executive Fellow at Henley Management College. He is also a visiting Tutor of Lille Graduate School of Management and Essex University. He specialises in Operational Excellence and Supply Chain Management and has research interests in Performance Management and Project Management.

Previously he held senior management roles in blue-chip companies like GSK, GlaxoWellcome and Unilever and led global initiatives and projects in Six Sigma, ERP/MRPII, Supply Chain Re-engineering and Total Productive Maintenance. Prior to this he worked as Management Consultant with A.T. Kearney.

He is the co-author of *Total Manufacturing Solutions*, *Quality Beyond Six Sigma, Total Operations Solutions* and *Total Supply Chain Management* and the author of books with titles *Measuring e-Business Performance* and *Implementing Quality.* He has authored a number of papers in the operational excellence and performance management fields. He is a regular presenter of papers in global seminars on e-Business, Six Sigma and Manufacturing and Supply Chain topics.

After graduating in Manufacturing Engineering from UMIST, Manchester, Ron obtained an MSc in Operational Research from Strathclyde University, Glasgow. He is a Fellow of the Institution of Mechanical Engineers, the Institute of Business Consultancy, the Association of Project Managers and the Chartered Quality Institute. He has also been selected as a Member of the International Who's Who for Professionals.

Part 1
The Foundations of Quality

Introduction to Chapters 1–3

Part 1 of the book deals with some of the fundamentals of quality. The perception of quality varies significantly depending on the type of industry, economy and culture. However it is undeniable that in today's global economy quality forms an integral part of the business, and the differences in its perception from the points of view of both the suppliers and customers are gradually converging. The following topics are examined in this section:

Chapter 1: Quality and operational excellence
Chapter 2: History of the quality movement
Chapter 3: The scope of tools and techniques

Chapter 1 addresses the dimensions of quality and emphasises that primarily the sustainability of quality can lead to operational excellence. Tools and techniques are essential aides in the road map to excellence.

Quality 'gurus', in particular Deming, Juran, Crosby, Feigenbaum, Taguchi and Ishikawa have had a significant influence in the development of the quality movement throughout the world and their learning and ideas are summarised in Chapter 2.

Chapter 3 examines the driving forces and opposing forces contributing to the success or failures of a quality programme in general and the application of tools and techniques in particular.

1

Quality and operational excellence

We are what we repeatedly do. Excellence, then, is not an act but a habit

— Aristotle

Introduction

The methodology of implementing a quality management and improvement programme can be varied. The programme is likely to have a different name or label, such as TQM (Total Quality Management), Six Sigma, Lean Sigma, BPR (Business Process Re-engineering) or Operational Excellence. Regardless of the methodology or name of the continuous improvement programmes, each organisation and programme team will certainly need to use a selection or tools and techniques in their implementation process. Most of these tools and techniques are simple to understand and can be used by a large population of the company. However, there are also some techniques which are more complex. These advanced techniques are used by specialists for specific problem solving applications. It is vital that the tools and techniques are selected for the appropriate team and applied correctly to the appropriate process. Therefore the fundamental requirements for achieving repeatable and reliable results by these tools and techniques is a clear understanding, both of the tools and techniques themselves and the process by which they could be applied.

The objective of this chapter is to introduce to the reader the following areas:

- What are the tools and techniques
- The concept of quality and operational excellence.

Tools and techniques

In general, tools and techniques can be broadly defined as the practical methods and skills applied to specific activities to enable improvement. A specific tool has a defined role and a technique may comprise the application of several such tools.

Dale and McQuater (1998) have suggested the following definition of tools and techniques.

Tools and techniques

A single tool may be described as a device which has a clear role and defined application. It is often narrow in its focus and can be and is usually used on its own. Examples of tools are:

- Cause and Effect Diagram
- Pareto Analysis
- Relationship Diagram
- Control Chart
- Histogram
- Flow Chart

A technique, on the other hand, has a wider application than a tool. There is also a need for a greater intellectual thought process and more skill, knowledge, understanding and training in order to use them effectively. A technique may even be viewed as a collection of tools. For example, Statistical Process Control employs a variety of the tools, such as graphs, charts, histograms and capability studies, as well as other statistical methods, all of which are necessary for the effective deployment of a technique. The use of a technique may cause the necessity for a tool to be identified.

Examples of techniques are:

- Statistical Process Control
- Benchmarking
- Quality Function Deployment
- Failure Mode and Effects Analysis
- Design of Experiments
- Self-assessment

Source: Dale and McQuater (1998).

What is quality?

If you were to ask quality experts to define 'quality', it is likely that you would receive many different answers, although you would elicit a set of common or comparable themes, such as 'Fitness for purpose', 'Right first time', 'What the customer wants', 'Conformance to standards', 'Value for money', 'Right thing at the right time' and so on. A basic reason for differing perceptions of quality is arguably that each person has their own set of individual preferences.

A simple story from the Indian fables may illustrate the above point. Three blind men went to visit an elephant and each felt the creature to form an impression of it. On their way back, they discussed the experience. The first man said, 'The elephant is just like a swinging fan'. The second blind man replied, 'No, I disagree. I think that it is more like a pillar'. Then the third person protested, 'You're both wrong. I would describe it as being more like a huge, thick whip.' He added, 'I am absolutely sure, it's a long and very flexible object'. It is clear from their very different impressions and viewpoints that the three blind men were influenced by their varying attitudes and the way in which they touched the elephant in order to arrive at such contrary perceptions about the same animal. However, it was only by sharing their ideas that they realised that they had visualised one concept in a variety of ways.

There are many different definitions and dimensions of quality to be found in books and academic literature. We will present three of these definitions selected from published literature and propose a three-dimensional definition of quality to reflect the appropriate application of tools and techniques.

One of the most respected definitions of quality is given by the eight quality dimensions (see Table 1.1) developed by David Gravin of the Harvard Business School (1984).

Table 1.1 Gravin's product quality dimensions

- Performance
- Features
- Reliability
- Conformance
- Durability
- Serviceability
- Aesthetics
- Perceived quality

Performance refers to the efficiency (e.g. return on investment) with which the product achieves its intended purpose.

Features are attributes that supplement the product's basic performance, e.g. tinted glass windows in a car.

Reliability refers to the capability of the product to perform consistently over its life cycle.

Conformance refers to meeting the specifications of the product, usually defined by numeric values.

Durability is the degree to which a product withstands stress without failure.

Serviceability is used to denote the ease of repair.

Aesthetics are sensory characteristics such as a look, sound, taste and smell.
Perceived quality is based upon customer opinion.

The above dimensions of quality are not mutually exclusive, although they relate primarily to the quality of the product. Neither they are exhaustive. Service quality is perhaps even more difficult to define than product quality. A set of service quality dimensions (see Table 1.2) that is widely cited has been compiled by Parasuraman et al. (1984).

Table 1.2 Parasuraman et al.'s service quality dimensions

- Tangibles
- Service reliability
- Responsiveness
- Assurance
- Empathy
- Availability
- Timeliness
- Professionalism
- Completeness
- Pleasantness

Tangibles are the physical appearance of the service facility and people.
Service reliability deals with the ability of the service provider to perform dependably.
Responsiveness is the willingness of the service provider to be prompt in delivering the service.
Assurance relates to the ability of the service provider to inspire trust and confidence.
Empathy refers to the ability of the service provider to demonstrate care and individual attention to the customer.
Availability is the ability to provide service at the right time and place.
Professionalism encompasses the impartial and ethical characteristics of the service provider.
Timeliness refers to the delivery of service within the agreed lead time.
Completeness addresses the delivery of the order in full.
Pleasantness simply means the good manners and politeness of the service provider.

Noriaki Kano (1996) demonstrates in the well-known Kano Model of customer satisfaction (see also Appendix A7) that there are three attributes to quality (viz. basic needs, performance needs and excitement needs) and that to be competitive products and services must flawlessly execute all three attributes of quality.

Our third authoritative definition of quality is taken from Ray Wild's *Operations Management* (2002, p. 644) (see Table 1.3).

Table 1.3 Wild's definition of quality

The quality of a product or service is the degree to which it satisfies customer requirements. It is influenced by: *Design quality*: The degree to which the *specification* of the product or service satisfies customers' requirements. *Process quality*: The degree to which the product or service, which is made available to the customer, *conforms* to specification.

The list of quality dimensions by Gravin (1984) and Parasuraman et al. (1985) are widely cited and respected. However, one problem with multiple dimensions is that of communication and if allowed time, the reader could probably identify additional dimensions. It is not easy to devise a strategic plan on quality based on specific dimensions which could be interpreted differently by different departments. Wild's definition of design/process quality however provides a broad framework to develop a company specific quality strategy.

Nonetheless, one important dimension of quality is not clearly visible in the above models: the quality of the organisation. This is a fundamental cornerstone of the quality of a holistic process and an essential requirement of an approved quality assessment scheme such as EFQM (European Foundation of Quality Management).

Our three-dimensional model of quality is shown in diagrammatic form in Figure 1.1.

Figure 1.1 Three dimensions of quality (© Ron Basu).

When an organisation develops and defines its quality strategy, it is important to share a common definition of quality and each department within a company can work towards a common objective. The product quality should contain defined attributes of both numeric specifications and perceived dimensions. The process quality, whether it relates to manufacturing or service operations, should also contain some defined criteria of acceptable service level so that the conformity of the output can be validated against these criteria. Perhaps the most important determinant of how we perceive sustainable quality is the functional and holistic role we fulfil within the organisation. It is only when an organisation begins to change its approach to a holistic culture emphasising a single set of numbers based on transparent measurement with senior management commitment that the 'organisation quality' germinates. We have compiled (see Table 1.4) a set of key organisation quality dimensions.

Table 1.4 Basu's organisation quality dimensions

- Top management commitment
- Sales and operations planning
- Single set of numbers
- Using tools and techniques
- Performance management
- Knowledge management
- Teamwork culture
- Self-assessment

Top management commitment means that organisational quality cannot exist without the total commitment of the top executive team.

Sales and operations planning is a monthly senior management review process to align strategic objectives with operation tasks.

Single set of numbers provides the common business data for all functions in the company.

Using tools and techniques relates to the fact that without the effective application of tools and techniques, the speed of improvement will not be assured.

Performance management includes the selection, measurement, monitoring and application of key performance Indicators.

Knowledge management includes education, training and development of employees, sharing of best practice and communication media.

Teamwork culture requires that teamwork should be practised in cross-functional teams to encourage a borderless organisation.

Self-assessment enables a regular health check of all aspects of the organisation against a checklist or accepted assessment process such as EFQM.

Hierarchy of quality

Our hierarchy of quality approximately follows the evolution of quality management from simple inspection to full quality management system as shown in Figure 1.2.

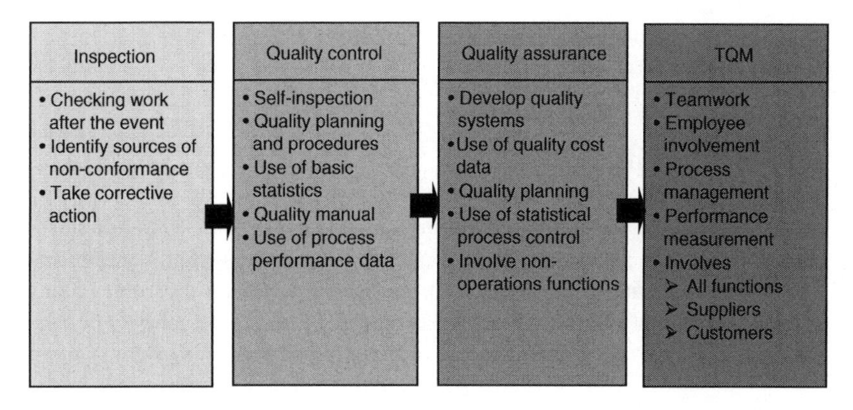

Figure 1.2 Hierarchy of quality.
(*Source*: Hill (2000); A similar approach is taken by Wright (1999), chapter 10 (this chapter is reprinted in the Managing Performance Readings).

Quality by inspection is an expensive method of achieving a basic level of quality. It requires the employment of people to check on the operation. Inspection and supervision do not add value to a product, they merely add to the cost! However the inspection of results with specified requirements are often necessary to ensure regulatory or approved standards.

Quality control (QC) is the next stage above quality inspection. The control process is based on the statistical method which includes the phases of analysis, relation and generalisation. Activities relating to QC include:

- Monitoring Process Performance
- Acceptance Sampling
- Designing and Maintaining Control Charts.

Quality assurance (QA) relates to activities needed to provide adequate confidence that an entity will fulfil requirements for quality. The first two stages, inspection and control, are based on a detection approach and relate to 'after the event', while QA is aimed at preventing mistakes. QA activities include:

– Approved Supplier Scheme
– Operator Training
– Process Improvement.

Total quality management has been defined in ISO 8402, 1995, as the 'Management approach of an organisation, centred on quality, based on the participation of all its members and aiming at long-term success through customer satisfaction, and benefits to all members of the organisation and society'. The holistic view of TQM supports the idea that quality is the responsibility of all employees and not just quality managers. TQM encompasses all three dimensions of quality as shown in Figure 1.1, with particular emphasis on organisational quality.

Cost of quality

One frequently asked question about quality management is, 'Can you quantify the benefits?' The answer to this question is yes we can, albeit approximately. The benefits are quantified in terms of not having the right quality or the cost of poor quality. As shown in Figure 1.3, cost of quality is derived

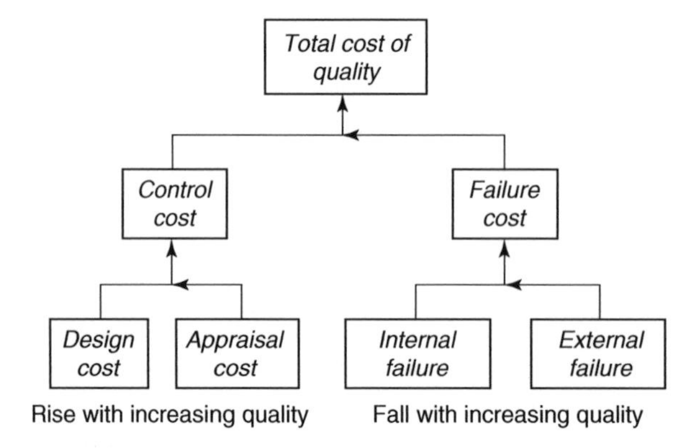

Figure 1.3 Cost of quality. (© Ron Basu; Also Wild's model for cost of quality. Text book, 6th edition, p. 650).

from the non-value added activities or wastes in the process and is made up of costs associated with

- Prevention
- Appraisal
- External failure
- Internal failure

The emphasis will be on prevention rather than detection, thus the cost of supervision and inspection will go down. Prevention will go up because of training and action-orientated efforts. But the real benefits will be gained by a significant reduction in failures – both internal (e.g. scrap, rework, downtime) and external (handling of complaints, servicing cost, loss of goodwill). The total cost of quality will reduce over time as shown in Figure 1.4 (see also Basu and Wright (1997, pp. 116–117)).

The concept of the law of diminished returns argues that there is a point at which investment in quality improvement will become uneconomical. Therefore the ethic of continuous improvement should aim at the appropriate or optimum level of quality and then sustain it.

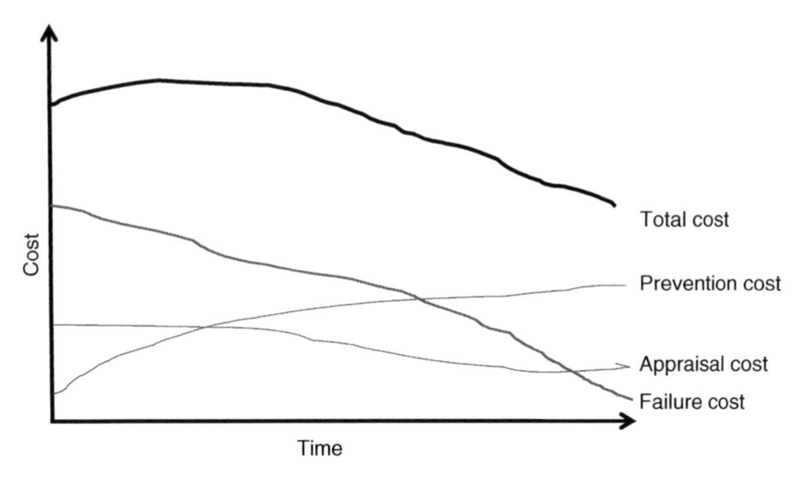

Figure 1.4 Total cost of quality (© Ron Basu).

Waves of quality management

Today, depending on whom you listen to, Six Sigma is either a revolution slashing trillions of dollars from Corporate inefficiency, or it's the most maddening management fad yet devised to keep front-line workers too busy collecting data to do their jobs.

— (*USA Today*, 21 July 1998)

It has been two and a half years since the above statement was made. During this time the 'Six Sigma revolution' has created a huge impact in the field of Operational Excellence, yet conflicting views are still prevalent.

Let us evaluate the arguments for both sides. On a positive note, the success of 'Six Sigma' in General Electric (GE) under the leadership of Jack Welch is undisputed. In the GE company report of 2000 their CEO was unstinting in his phrase: 'Six Sigma has galvanised our company with an intensity the likes of which I have never seen in my 40 years of GE'. Even financial analysts and investment bankers compliment the success of Six Sigma in GE. An analyst at Morgan Stanley Dean Witter recently estimated that GE's gross annual benefit from Six Sigma could reach 5% of sales and that share value might increase by between 10% and 15%.

However the situation is more complex than such predictions would suggest.

In spite of the demonstrated benefits of many improvement techniques such as TQM, BPR and Six Sigma, most attempts by companies to use them have ended in failure (Easton and Jarrell, 1998). Sterman et al. (1999) conclude that companies have found it extremely difficult to sustain even initially successful process improvement initiatives. Yet more puzzling is the fact that successful improvement programmes have sometimes led to declining business performance causing lay offs and low employee morale. Motorola, the originator of Six Sigma, announced in 1998 that its second quarter profit was almost non-existent and that consequently it was cutting 15 000 of its 150 000 jobs!

To counter heavyweight enthusiasts like Jack Welch (GE) and Larry Bossidy (Allied Signal) there are sharp critics of Six Sigma. Six Sigma may sound new, but critics say that it is really Statistical Process Control in new clothing. Others dismiss it as another transitory management fad that will soon pass.

It is evident that like any good product 'Six Sigma' should also have a finite life cycle. In addition, Business Managers can be forgiven if they are often confused by the grey areas of distinction between quality initiatives such as TQM, Six Sigma and Lean Sigma.

Against this background, let us examine the evolution of total quality improvement processes (or in a broader sense Operational Excellence) from Ad hoc Improvement to TQM to Six Sigma to Lean Sigma. Building on the success factors of these processes the key question is: How do we sustain the results? The author has named this sustainable process as FIT SIGMA™ (see Basu and Wright, 2003).

What is FIT SIGMA™? Firstly, take the key ingredient of quality, then add accuracy in the order of 3.4 defects in 1 000 000. Now implement this across your business with an intensive education and training programme. The result is Six Sigma. Now let's look at Lean Enterprise, an updated version of classical Industrial Engineering. It focuses on delivered value from a customer's perspective and strives to eliminate all non-value added activities ('waste') for each product or service along a value chain. The integration of the complementary approaches of Six Sigma and Lean Enterprise is known as Lean Sigma. FIT SIGMA™ is the next wave. If Lean Sigma provides agility and efficiency, then FIT SIGMA™ allows a sustainable fitness. In addition the control of variation from the mean (small Sigma 'σ') in the Six Sigma process is transformed to company-wide integration (capital Sigma 'Σ') in the FIT SIGMA™ process. Furthermore, the philosophy of FIT SIGMA™ should ensure that it is indeed fit for the organisation.

The road map to FIT SIGMA™ (see Figure 1.5) contains three waves and the entry point of each organisation will vary:

First Wave: As Is to TQM
Second Wave: TQM to Lean Sigma
Third Wave: Lean Sigma to FIT SIGMA™

First Wave: As Is to TQM

The organised division of labour to improve operations may have started with Adam Smith in 1776. However, it is often the Industrial Engineering approach, which has roots in F.W. Taylor's[1] 'Scientific Management (1929)',

[1]Frederick W. Taylor a late 19th century American is remembered as the father of scientific management. His philosophy was that management, by scientific means, should find the best method of doing a job (method and equipment). Once the best method was found workers were trained and offered incentives to increase productivity. Supervisors were employed to maintain the best method. Workers were not expected to make suggestions; their job was to do what they were told while management did the thinking.

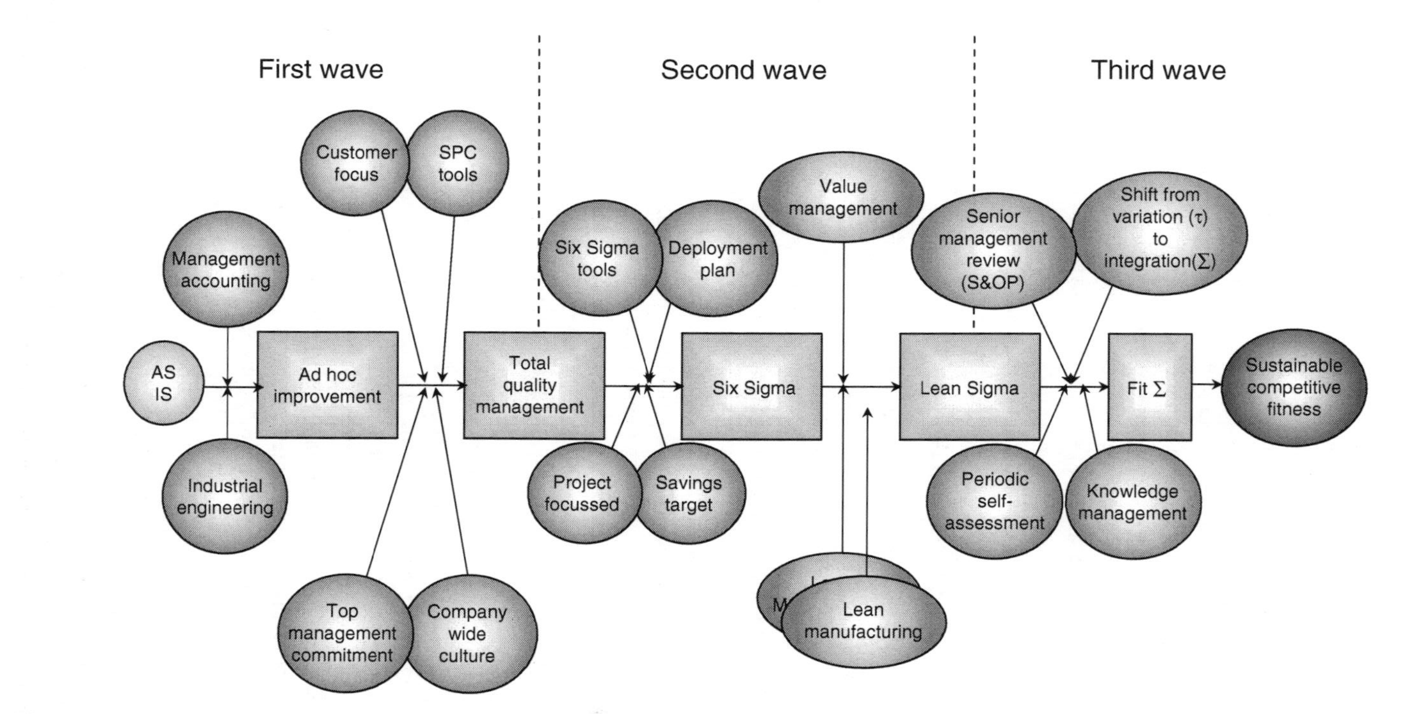

Figure 1.5 Road map to Operational Excellence (© Ron Basu).

that is credited with the formal initiation of the first wave of Operational Excellence. This Industrial Engineering approach was sharpened by Operational Research and complemented by operational tools such as Management Accounting.

During the years following the Second World War, the 'first wave' saw through the rapid growth of industrialisation, but in the short term the focus seemed to be upon both increasing volume and reducing the cost. In general, improvement processes were 'ad hoc', factory centric and conducive to 'pockets of excellence'. Then in the 1970s the holistic approach of TQM initiated the 'second wave' of Operational Excellence. The traditional factors of QC and QA are aimed at achieving an agreed and consistent level of quality. However TQM goes far beyond mere conformity to standard. TQM is a company-wide programme and requires a culture in which every member of the organisation believes that not a single day should go by within the organisation without in some way improving the quality of its goods and services.

Second Wave: TQM to Lean Sigma

Learning the basics from W.E. Demming and J.M. Juran Japanese companies extended and customised the integrated approach and culture of TQM (Basu and Wright, 1997). Arguably the economic growth and manufacturing dominance of Japanese industries in the 1980s can be attributed to the successful application of TQM in Japan. The three fundamental tenets of Juran's TQM process are firstly, upper management leadership of quality; secondly, continuous education on quality for all; finally, an annual plan for quality improvement and cost reduction. These foundations are still valid today and embedded within the Six Sigma/Lean Sigma philosophies. Phil Crosby and other leading TQM Consultants incorporated customer focus and Demming's SPC tools and propagated the TQM philosophy to the USA and the industrialised world. The Malcolm Baldridge Quality Award, ISO 9000 and Demming Quality Award have enhanced the popularity of TQM throughout the world, while in Europe the EFQM was formed. During the 1980s TQM seemed to be everywhere and some of its definitions such as 'fitness for the purpose', 'quality is what customer wants' and 'getting it right first time' became so over used that they were almost clichés. Thus the impact of TQM then began to diminish.

In order to complement the gaps of TQM in specific areas of Operation Excellence high-profile consultants marketed mostly Japanese practices in the form of a host of three letter acronyms (TLAs) such as JIT, TPM, BPR and MRPII. Total Productive Maintenance (TPM) has demonstrated successes outside Japan by focusing on increasing the capacity of individual processes. TQM was the buzzword of the 1980s but it is viewed by many, especially in the US quality field, as an embarrassing failure – a quality concept that promised more than it could deliver. Philip Crossby pinpoints the cause of TQM 'failures' as 'TQM never did anything to define quality, which is conformance to standards. Perhaps the pendulum swung too far towards the concept of quality as 'goodness' and employee culture. It was against this background that the scene for Six Sigma appeared to establish itself.

Six Sigma began back in 1985 when Bill Smith, an engineer at Motorola, came up with the idea of inserting hard nosed statistics into the blurred philosophy of quality. In statistical terms, Sigma (σ) is a measure of variation from the mean and the greater the value of Sigma the fewer the defects. Most companies produce result at best around four Sigma or more than 6000 defects. By contrast at the Six Sigma level, the expectation is only 3.4 defects per million as companies move towards this higher level of performance.

Although invented in Motorola, Six Sigma has been experimented with by Allied Signal and perfected at GE. Following the recent merger of these two companies, GE is truly the home of Six Sigma. During the last 5 years, Six Sigma has taken the quantum leap into Operational Excellence in many blue chip companies including DuPont, Ratheon, Ivensys, Marconi, Bombardier Shorts, Seagate Technology and GlaxoSmithKline.

The key success factors differentiating Six Sigma from TQM are:

- The emphasis on statistical science and measurement.
- A rigorous and structured training deployment plan (Champion, Master Black Belt, Black Belt and Green Belt).
- A project focused approach with a single set of problem solving techniques such as DMAIC (Define, Measure, Analyse, Improve, Control).
- Reinforcement of Juran tenets (Top Management Leadership, Continuous Education and Annual Savings Plan).

Following their recent application in companies like GlaxoSmithKline, Ratheon, Ivensys and Seagate, the Six Sigma programmes have moved to the Lean Sigma philosophy, which integrates Six Sigma with the complementary approach of Lean Enterprise. Lean focuses the company's resources and its suppliers on the delivered value from the customer's perspective. Lean Enterprise begins with Lean production, the concept of waste reduction developed from industrial engineering principles and refined by Toyota. It expands upon these principles to engage all support partners and customers along the value stream. Common goals to both Six Sigma and Lean Sigma are the elimination waste and improvement of process capability. The industrial engineering tools of Lean Enterprise complement the science of the statistical processes of Six Sigma. It is the integration on these tools in Lean Sigma that provides an Operational Excellence methodology that addresses the entire value delivery system.

Third Wave: Lean Sigma to FIT SIGMA™

Lean Sigma is the beginning of the 'third wave'. The predictable Six Sigma precisions combined with the speed and agility of Lean produces definitive solutions for better, faster and cheaper business processes. Through the systematic identification and eradication of non-value added activities, optimum value flow is achieved, cycle times are reduced and defects eliminated.

The dramatic bottom-line results and extensive training deployment of Six Sigma and Lean Sigma must be sustained with additional features for

---

FIT SIGMA™ is a registered trademark of Performance Excellence Limited, UK.

securing the longer-term competitive advantage of a company. The process to do just that is FIT SIGMA™. The best practices of Six Sigma, Lean Sigma and other proven Operational Excellence best practices underpin the basic building blocks of FIT SIGMA™.

Four additional features are embedded in the Lean Sigma philosophy to create FIT SIGMA™. These are:

1. A formal Senior Management Review process at regular intervals, similar to the Sales and Operational planning process.
2. Periodic self-assessment with a structured checklist which is formalised by a certification or award, similar to EFQM award but with more emphasis on self-assessment.
3. A continuous learning and knowledge management programme.
4. The extension of the programme across the whole business with the shifting of the theme of variation control (σ) of Six Sigma to the integration of a seamless organisation (Σ).

Please refer to separate illustration in Figure 1.5.

Operational Excellence

In agreement with the waves of quality management, we have developed a road map towards Operational Excellence and we note a convergence between the two. Operational Excellence (OE) is a broader programme of improving and sustaining business performance in which quality management is embedded. OE is synonymous with Business Excellence and it also encompasses other focused excellence programmes such as Manufacturing Excellence, Service Excellence, Marketing Excellence and Supply Chain Excellence. For most companies to be the very best in their industry and stay there is a long journey which will need a long-term improvement plan of, say, 5–10 years.

Figure 1.6 helps us ascertain the steps towards achieving Operational Excellence.

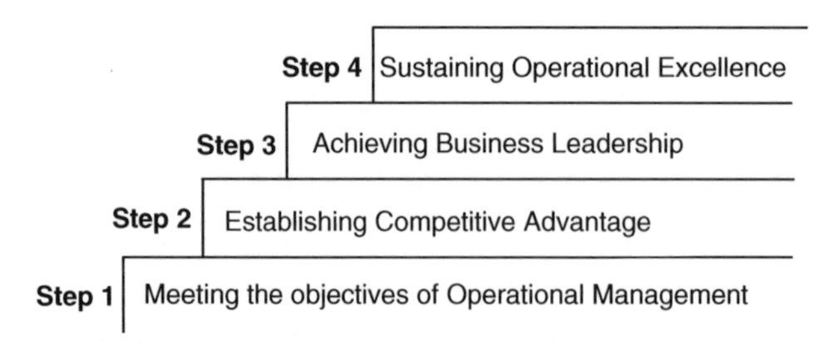

- Both continuous improvement and step changes are essential ingredients of Operational /Excellence
- OE means best managed business

Figure 1.6 Steps of Operational Excellence (© Ron Basu).

Step 1: This is the stage of operational management where the business objective is to keep it profitable by balancing what the customer wants with the resources available.

Step 2: The company aims to achieve competitive advantage by benchmarking with competitors. This is supported by a continuous improvement programme. We call this stage Operational Improvement.

Step 3: The company's aim is to be 'best in class' and the strategy is to deploy a well-managed holistic quality programme such as TQM or Six Sigma. This is the stage of OE.

Step 4: The aim of Step 4 is to sustain the benefits of Step 3 or OE. The strategy will include sustainable processes (e.g. self-assessment, knowledge management) of FIT SIGMA™ (see Basu and Wright, 2003).

It is evident that both continuous improvement and step changes are integral ingredients of OE.

2

History of the quality movement

There is nothing new under the sun. Is there a thing of which it is said, 'See, this is new?' It has been already in the ages before us.

— Ecclesiastes, Old Testament

Introduction

Over the last three decades high profile consultants have marketed quality initiatives in the form of a host of three letter acronyms (TLAs), such as TQM, JIT, BPR, TPM, MRP2, ERP, etc. The consultants are careful to point out that TLAs are not 'quick fixes'. But by the very act of this warning they subtly imply that, if their firm is consulted, a quick fix can happen. There are some success stories but there are also many failures. Further one consulting firm often derides the previous initiative to promote their own 'solutions' in search of problems. Therefore it is not surprising that a CEO may draw little comfort from the writings of quality gurus in order to select an appropriate quality programme in the organisation. There little doubt that managers are confused by the variety of advice available. In order to make the tools and techniques effective the organisation needs a quality programme and the selection of the programme is underpinned by a broad understanding of both the distinctive and common features of each initiative.

The Malcom Baldridge award, the Deming quality award, The European Foundation of Quality, have all served to give TQM a high profile. One count suggests that there are over 400 TQM tools and techniques (Pyzdek, 2000). However, this high profile has paradoxically contributed to a level of scepticism, especially by middle managers and staff. Promises have not been realised, high profile organisations that have claimed to be practising TQM have gone into decline, and staff have seen slogans and mission statements published which focus on customer service and people coupled with TQM followed by redundancies and drastic cuts in training budgets. This chapter discusses what is meant by quality and gives a historical overview of the development of quality thinking beginning with Total Quality Management (TQM). It concludes with a summary of how FIT SIGMA™ builds on prior quality initiatives.

World class

The term 'world class' is generally attributed to Hayes and Wheelwright (1984), who related best practice to German and Japanese firms competing in export markets. Schonberger (1986) used the term best practice to describe manufacturers making rapid and continuous improvement. World class in the 1990s was extended to include lean production (see Womack et al. (1990) referred to in chapter 1).

Fry et al. (1994) and Harrison (1998) say best practice refers to any organisation that performs as well or better than the competition in quality, timeliness, flexibility and innovation. Knuckey et al. (1999, p. 23) explain that 'the logic behind best practice is simple: because operational outcomes are a key contributor to competitiveness and business performance, and because best practice should improve operational outcomes, by implication good practice should lead to increased competitiveness. Best practice should lead to world class service'.

Knuckey et al. (1999, p. 137), on behalf of the New Zealand Ministry of Commerce, found from research of 1173 New Zealand manufacturing firms that the 'main sources of competitive advantage' and 'best practice' is:

goodwill and trust with suppliers and distributors,
trust, goodwill and commitment from employees to the firm's goals, and
reputation with clients.

Why best practice and world class is essential

There is no doubt that people today are more travelled, better educated and consequently more discerning than ever before. Customers know what is on offer elsewhere, they have experienced it and their expectations have been raised by advertising and marketing. Likewise shareholders, and other financial stakeholders, can be excused for wondering why the rapid technological advances of the last decade have not resulted in increased performance and higher returns on investment. At the same time the well publicised and promised benefits of technology have led customers to expect, even demand, improved products and service at less cost. Quality service, reliable products, value for money and accountability are now taken for granted. Competitors are global, standards are world class and organisations that fail to meet world-class performance will soon be found out.

W. Edwards Deming

TQM has its origins in Japan. In the 1960s Japan went through a quality revolution. Prior to this 'Made in Japan' meant cheap or shoddy consumer goods. The approach used in Japan in the 1950s and 1960s to improve quality standards was to employ consultants from America and Europe. The most famous of these consultants was Dr W Edwards Deming. Deming's philosophy was to establish the best current practices within an organisation, establish the best practice as standard procedure and train the workers in the best way. In this manner, everyone would be using the same best way. Deming's approach was to involve everyone in the organisation and to win them over. He believed that quality was everyone's business. Deming, W.E. (1986) said that to find the best way meant getting the facts, collecting data, setting standard procedures, measuring results, and getting prompt and accurate feedback of results so as to eliminate variations to the standard. He saw this as a continuous cycle. He claimed that cultivating the know-how of employees was 98% of the quality challenge – as Gabor (2000) says Deming has been criticised for hyperbole. No section on Dr Deming is complete without reference to his famous 14 points of quality. Comments in italics are our notes and not direct quotes of Deming.

1. Create consistency of purpose towards improvement of product and service.
2. Adopt the new philosophy; *management has to learn its responsibilities and to take leadership. It is difficult for management to accept that 90% of problems lie with management and the process.*
3. Cease dependence on inspection to achieve quality *supervision and supervisors wages do* not *add value, they are an extra cost, far better if staff take responsibility and supervise themselves. Deming also added that if quality is built into the design or process, then inspection will not be necessary.*
4. End the practice of awarding business on the basis of the price tag. *The cheaper the price the higher the number of failures. Move to dedicated suppliers and value reliability, delivery on time and quality.*
5. Improve constantly and forever the system of production and service. *This is an extension of the Japanese philosophy of Kaizen whereby not a day should go by without some incremental improvement within the organisation.*
6. Institute training on the job. *Become a learning organisation with a willingness to share knowledge.*
7. Institute leadership, everyone, *at all levels, especially supervisors to be team leaders not disciplinarians. Everyone to be encouraged to develop self-leadership. Quality is too important to be left to management.*
8. Drive out fear. *Encourage people to admit mistakes. The aim is to fix not to punish. However it is expected that people won't go on making the same mistakes!*
9. Break down barriers between departments, *eliminate suspicion between departments. There needs to be clear objectives with everyone striving to work for the common good.*
10. Eliminate slogans, exhortations and targets for the workforce. *There is no use asking for zero defects if the process or the product design is not perfect. Ten per cent across the board cost reduction demands are poor for morale if they are not possible.*

11. Eliminate work standards – quotas – on the factory floor. *For example, 100 pieces per hour with a bonus for a 110 will result in 110 pieces, but not necessarily quality products. The focus will be on output numbers rather than quality. If the worker is encouraged to consider quality, 95 high quality pieces per hour will be worth more than 110 if 15 (of the 110) are subsequently rejected or returned by the customer.*
12. (a) Remove barriers that rob the worker of the right to pride of workmanship *give them the right tools, right materials, right processes and comfortable working conditions, treat them with respect.*
    (b) Remove barriers that rob people in management and engineering of their right to pride in craftsmanship. *This includes appraisal systems which reward on bottom line results and keeping expense budgets Dow, and ignores customer satisfaction. If cost is the only driver, then training, maintenance, and customer service, etc. will suffer.*
13. Institute a programme of education and self-improvement. *Encourage staff to seek higher educational qualifications. Become a knowledge based organisation.*
14. Put everybody in the company to work to accomplish the transformation. *Change of culture is difficult to achieve. Dr Deming saw that everyone has to be involved in transforming the culture of an organisation.*

Deming was not the only guru of quality used by the Japanese.

Joseph M. Juran

Dr Joseph M Juran was also associated with Japan's emergence as the benchmark for quality of products. Like Deming, Juran was an American statistician and there are similarities between his work and that of Deming. Above all both men highlight managerial responsibility for quality. Arguably Juran was the first guru to emphasise quality was achieved by communication. The Juran (1989) trilogy for quality is, planning, control and improvement. His approach includes an annual plan for quality improvement and cost reduction, and continuous education on quality. Juran's foundations are still valid and are embedded within Six Sigma and Lean Sigma and our FIT SIGMA™ philosophies. He argues, and few would disagree, that inspection at the end of the line, post-production, is too late to prevent errors. Juran says that quality monitoring needs to be performed during the production process to ensure that mistakes do not occur and that the system is operating effectively. Juran does this by examining the relationship between the process variables and the resultant product. Once these relationships have been determined by statistical experiment the process variables can be monitored using statistical methods. Juran adds that upper management role is more than making policies; they have to show leadership through action, they have to walk the talk, not just give orders and set targets. Juran says quality is not free and that investment is needed (often substantial investment) in training, often including statistical analysis, at all levels of the organisation. Juran also believed in the use of Quality Circles. As he describes them, Quality Circles are small teams of staff with a common interest who are brought together to solve quality problems. Our constituents for a successful Quality Circle are discussed later in this chapter.

Armand V. Feigenbaum

Feigenbaum, A.V. (1983) is recognised for his work in raising quality awareness in the United States. He was General Electric's worldwide chief of manufacturing operations for a decade until the late 1960s. The term Total Quality Management originated from his book *Total Quality Control* (1961). Feigenbaum states that Total Quality Control has an organisation wide impact which involves managerial and technical implementation of customer-orientated quality activities as a prime responsibility of general management and of the main line operations of marketing, engineering, production, industrial relations, finance and service as well as of the quality-control function itself. He adds that a quality system is the agreed, company wide operating work structure, documented in integrated technical and managerial procedures, for guiding the coordinated actions of the people, the machines and company wide communication in the most practical ways with the focus on customer quality satisfaction.

Feigenbaum is one of the first writers to recognise that quality must be determined from the customers perspective, and NOT the designer's (or the engineer's or the marketing department's concept of what quality is).

Feigenbaum also says that the best does not mean outright excellence, but means best for satisfying certain customer conditions. In other words, like FIT SIGMA™, best means sufficiently good to meet the circumstances. Feigenbaum, also like Deming and Juran, found that measurement is necessary. But, whereas Dening and Juran tended to measure production and outputs, Feigenbaum concentrated on measurement to evaluate if good service and product met the desired level of customer satisfaction.

According to Dale (1999), Feigenbaum's major contribution to quality was to recognise that the three major categories of cost are: appraisal, prevention and cost of failure. According to Feigenbaum the goal of quality improvement is to reduce the total cost of quality from the often quoted 25–30% of cost of sales (a huge per cent when you think about it) to as low a per cent as possible. Thus Feigenbaum takes a very financial approach to the cost of quality.

Philip B. Crosby

Philip B Crosby a guru of the late 1970s was the populist who 'sold' the concept of TQM and 'zero defects' to the United States (Crosby, P.B. (1979a)). Although zero defects sounds very much like Six Sigma, in fact Crosby takes a very much softer approach than does Deming, Juran, Feigenbaum or Six Sigma. His concept of zero defects is based on the assumption that it is always cheaper to do things right the first time and quality is conformance to requirements. Note the wording 'conformance to requirements'. Thus any product that conforms to requirements, even where requirements are specified at less than perfection, would be deemed to be defect free.

Crosby developed the concept of non-conformance when recording the cost of quality. Non-conformance includes the cost of waste and scrap, down time due to poor maintenance, putting things right, product recall, replacement and at worst legal costs. All these can be measured, and according to Crosby cost of non-conformance 'can be as much as 20% of manufacturing sales and 30% of operating costs in service industries'.

Crosby is famous for saying quality is free. And he wrote a book with that title (Crosby, 1979b). He emphasised cultural and behavioural issues ahead of the statistical approach of Deming and Feigenbaum. Crosby was saying, if staff have the right attitude, know what the standards are and do things right first time every time, the cost of conformance will be free. The flow on effect is that motivated workers will go further than just doing things right, they will detect problems in advance, they will be proactive in correcting situations and they will be quick to suggest improvements. Crosby concluded workers should not be blamed for errors, but rather that management should take the lead and the workers will follow. Crosby suggests that 85% of quality problems are within managements control (Deming put this figure at 90%).

Hammer and Champy

The concept of Business Process Re-engineering (BPR) was popularised by Hammer and Champy (1993) and they defined BPR as 'the fundamental rethinking and radical redesign of business processes to achieve dramatic improvements in critical contemporary measures performance'. According to Hammer and Champy BPR is not something that can be accomplished in gradual steps like TQM. Rather a 'big bang' approach that in theory sets aside any pre-conceived ideas of the structures and processes of an organisation. Hammer and Champy also claims that many companies still adhere to the nearly 200-year-old Adam Smith concept of work structure.

Hammer and Champy introduced popular models, the 'five roles' which were seen to be essential to implement a BPR programme:

1. *The Leader*: a senior executive who authorises and champions the overall programme.
2. *The Process owner*: a manager with responsibility for a specific process re-engineering.
3. *The Re-engineering Team*: a group of individuals dedicated to the re-engineering of a particular process.
4. *The Re-engineering Committee*: a policy-making body of senior managers who develop the organisation's overall re-engineering strategy and monitor its progress.

5. *The Re-engineering Czar*: an individual responsible for developing re-engineering tools and techniques for achieving synergy across separate projects.

BPR became popular in early 1990s driven by the information technology (IT). While IT has been very useful in these efforts, it often proved to serve as the 'tail wagging the dog'. On the downside many BPR efforts (by some estimates over 50%) failed or at best yield only marginal improvements. According to Dale (1999), 'the underlying issues in BPR are not necessarily new, albeit the language is modern.' Perhaps it is fair to point out that the concept of process redesign earlier by Harrington (1987) and Davenport and Short (1990) was the origin of BPR. The characteristics of BPR have often been compared with those of TQM. Oakland (2003) considers the distinct, short-term activities of BPR are complementary to the longer-term objectives of TQM. According to Oakland, BPR has been 'a process-driven change dedicated to the ideals and concepts of TQM'. Davenport (1993) summarised the features of contrast between BPR and TQM as shown in Table 2.1.

Table 2.1 BPR versus TQM

	TQM	BPR
Level of change	Incremental	Radical
Starting point	Existing process	Clean slate
Frequency of change	Continuous	One-time
Participation	Bottom up	Top down
Risk	Moderate	High
Primary enabler	Statistical process control	Information technology
Type of change	Cultural	Structural

What of the Japanese?

The most important of the Japanese writers on quality are Genichi Taguchi, Ishikawa, Shingo and Imai. And of course Toyota is widely cited as the epitome of lean production (see Womack et al. 1990).

Of the Japanese approaches to quality, Taguchi methods have been the most widely adopted in America and Europe. Taguchi, an electrical engineer, used an experimental technique to assess the impact of many parameters on a single output. His method was developed during his work of rebuilding the Japanese telephone system in the 1970s. His approach to quality control is focused on 'off line' or loss of function (derived from telephone system failures).

The Taguchi approach is to:

- Determine the existing quality level measured in the incidence of down time – which he called 'off line'.
- Improve the quality level by parameter and tolerance design.

- Monitor the quality level by using statistical process control (SPC), to show upper and lower level variances.

Taguchi advocates three stages of quality design, namely;

1. *System design*: The development of the basic system which involves experimentation with materials, and the testing of feasibility with prototypes. Obviously technical/scientific knowledge is a requisite.
2. *Parameter design*: This begins with establishing the optimum levels for control factors so that the product or process is least sensitive to the effect in changes of conditions (i.e. the system is robust). This stage includes experimentation with the emphasis on using low cost materials and processes.
3. *Tolerance design*: This includes setting numerical values (factors) for upper service levels and lower acceptable service levels and reconciling the choice of factors in product design. This includes comparison of costs by experimenting with low cost materials and consideration of more expensive materials to reduce the tolerance gap. Design includes process design and product design. Process design includes choosing the upper and lower parameters of service, and product design includes reconciling the choice of materials against the desired service level parameters.

Taguchi promotes three stages in developing quality in the design of product or systems. They are:

1. Determine the quality level, as expressed in his loss function concept.
2. Improve the quality level in a cost effective manner by parameter and tolerance design.
3. Monitor the quality of performance by use of feedback and statistical control.

Lean Enterprise

Ohno Taiichi, after visiting USA car manufacturers in the 1960s, returned to Japan and developed a new method of manufacturing, which became known as lean production. Lean is more than a system it is a philosophy, began with Japanese automobile manufacturing in the 1960s, and was popularised by Womack et al. (1990) in *The Machine that Changed the World. The Machine that Changed the World* is essentially the story of the Toyota way of manufacturing automobiles. The characteristics of Lean, sometimes referred to as Toyotaism, are that materials flow 'like water' from the supplier through the production process onto the customer with little if any stock of raw materials or components in warehouses, no buffer stocks of materials and part finished goods between stages of the manufacturing process and no output stock of finished goods. This 'just in time' (JIT) approach requires that materials arrive from dedicated suppliers on the factory floor at the right stage of production just when required, and when the production process is completed it is shipped directly to the customer. With no spare or safety stock in the system there is no room for error. Scheduling of activities and resource has to be exact, communication with suppliers must be precise, suppliers have to be reliable and able to perform to exacting time tables, materials have to arrive on time and meet the specification, machines have to be maintained so that there is no down time, operators cannot make mistakes, there is no allowance for scrap or rework and finally the finished product has to be delivered on time to customers. This is often implemented by circulating cards or Kanbans between a work station and the downstream buffer. The work station must have a card before it can start an operation. It can pick raw materials out of its upstream (or input) buffer, perform the operation, attach the card to the finished part and put it in to the downstream (or output) buffer. The card is circulated back to the upstream to signal the next upstream work station to do next cycle. The number of cards circulating determines the total buffer size. Kanban control ensures that parts are made only in response to a demand. With computer controlled production, the Kanban principle applies but there is not a physical movement of cards, information is transferred electronically.

This 'JIT' approach generally precludes large batch production, instead items are made in 'batches' of one. This means that operators have to be flexible, the system has to be flexible and 'single minute exchange of dies' (SMED) becomes the norm. A lean approach reduces the number of supervisors and quality inspectors. The operators are trained to know the production standards required and are authorised to take corrective action, in short they become their own inspectors/supervisors.

The original Toyota model of Lean Manufacturing, from which various hybrids were developed, comprised eight tools and approaches:

1. TPM (Total Productive Maintenance): see Basu and Wright (1998), pp. 96–99.
2. 5Ss: These represent a set of Japanese words for excellent house keeping (Sein – Sort, Seiton – Set in place, Seiso – Shine, Seiketso – Standardise and Sitsuke – Sustain).
3. JIT (Just in Time).
4. SMED (Single Minute Exchange of Dies).
5. Judoka or Zero Quality Control.
6. Production Work Cells.
7. Kanban (see above).
8. Poka Yoke.

These terms, and others, are explained in various sections and also in the glossary at the end of the book.

Total Productive Maintenance

TPM includes more than maintenance, it addresses all aspects of manufacturing performance. The two primary goals of TPM are to develop optimum conditions for the factory through a self-help people/machine systems culture and to improve the overall quality of the workplace. It involves every employee in the factory. Since mid-1980s TPM has been promoted throughout the world by the Japan Institute of Plant Maintenance (JIPM). TPM is based on five key principles or 'pillars':

1. The improvement of manufacturing efficiency by the elimination of 'six big losses'.
2. The establishment of a system of autonomous maintenance by operators working in small groups.
3. An effective planned maintenance system by expert engineers.
4. A training system to increase the skill and knowledge level of all permanent employees.
5. A system of maintenance prevention where engineers work closely with suppliers to specify and design equipment which requires less maintenance.

TPM has been applied both in Japan and outside Japan in three ways: as a stand-alone programme, as part of Lean Manufacturing and as the manufacturing arm of TQM. More information on TPM can be found in Basu and Wright (1998), chapter 5, and Shirose (1992).

Basu and Wright and TQM

Basu and Wright (1998) identify a hierarchy of quality management which has four levels, which is further supported by Hill (2000).

Their four levels of quality management are: Inspection, Control, Assurance and TQM.

Quality inspection and quality control rely on supervision to make sure that no mistakes are made. The most basic approach to quality is inspection and detection and correction of errors. The next level, quality control, is to inspect, correct, investigate and find the causes of problems and to take actions to prevent errors re-occurring. Both methods rely on supervision and inspection. The third level, quality assurance, includes the setting of standards with documentation and also includes the documentation of the method of checking against the specified standards. Quality assurance generally also includes a third party approval from a recognised authority, such as found with the International Standard Organisation (ISO) 9000 series. With quality assurance, inspection and control are still the basic approach, but in addition one would also expect to find a comprehensive quality manual, the recording of quality costs, and perhaps use of SPC and the use of sampling techniques for random and the overall auditing of quality systems.

Quality inspection and control and quality assurance are aimed at achieving an agreed consistent level of quality firstly by testing and inspection, then by rigid conformance to standards and procedures, and finally by efforts to eliminate causes of errors so that the defined accepted level will be achieved.

TQM is on a different plane. TQM does, of course, include all the previous levels of setting standards and the means of measuring conformance to standards. In doing this, SPC may be used, systems will be documented and accurate and timely feedback of results will be given. With TQM, ISO accreditation might be sought, but an organisation that truly has embraced TQM does not need the ISO stamp of approval. ISO is briefly discussed later in this chapter.

Any organisation aspiring to TQM will have a vision of quality which goes far beyond mere conformance to a standard. TQM requires a culture whereby every member of the organisation believes that not one day should go by without the organisation in some way improving the quality of its goods and services. The vision of TQM must begin with the chief executive. If the chief executive does not have a passion for quality and continuous improvement, and if this passion can not be transmitted down through the organisation, then, paradoxically, the ongoing driving force will be from the bottom up.

Figure 2.1 depicts a TQM culture, wherein management has the vision, which is communicated to and accepted by all levels of the organisation. Once the quality culture has been ingrained in the organisation the ongoing driving force is 'bottom up'.

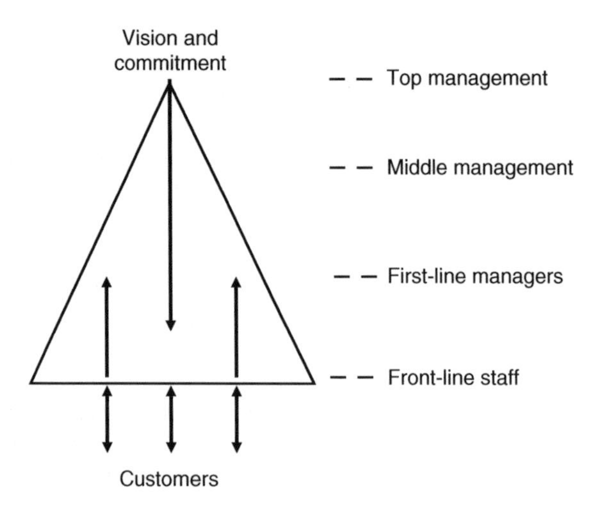

Figure 2.1 Quality and the driving force (© Ron Basu).

Once the culture of quality has become ingrained, quality will be driven from bottom up be it the factory worker, or the sales assistant rather than achieved by direction or control from the top.

Management will naturally have to continue to be responsible for planning and for providing the resources to enable the workers to do the job. But, unless the machine operator, the shop assistant, the telephone operator, the cleaner, the van driver and the junior account clerk are fully committed to quality, TQM will never happen.

TQM also goes beyond the staff of the organisation; it involves suppliers, customers and the general public.

Basu (2001) also introduced the concept of FIT SIGMA™ as a total business philosophy of process, performance and culture. The concept and its applications were explained in more details by Basu and Wright (2003). FIT SIGMA™ is the process that enables the dramatic bottom line results of Six and Lean Sigma to be sustained. It ensures where extensive training and development of skilled Sigma practitioners (Master, Black Belt and Green Belt) has been carried out, it is not wasted and the benefits are secured for the long term.

FIT SIGMA™ adds the following additional features to Six and Lean Sigma.

- A formal senior management review at regular intervals, similar to the sales and operational planning process.
- Periodic self-assessment with a structured checklist, which is recognised by a certificate or award, similar to the European Foundation of Quality Management or Baldridge process.
- A continuous learning and management program.
- A whole systems approach across the whole organisation.

American and European approaches to TQM

We have discussed the Japanese overall approach to TQM in our sections relating to W.E. Deming, Genichi Taguchi and others. As we have seen, the approaches that have emanated in one culture, such as the Taguchi method have crossed national boundaries, and likewise the principles of TQM as practiced in Japan were picked up by Feigenbaum and Crosby in the United States (and rapidly spread through Europe, the subcontinent of India and the rest of the world).

In the United States and in Europe, senior management, bankers and investors have a morbid fascination with share prices. They consequently feel pressure to meet short-term 6-monthly targets of interim and annual reports which are widely publicised and scrutinised. On each occasion a report or statement is made they must show a healthy bottom line, or at least provide a promise of better results in the short term for seeable future. The Japanese, however, know that success is rarely an overnight phenomenon.

The implementation of TQM, because it requires a total change in management thinking and a major change in culture, will take years to internalise. Thus with some organisations, because results are not instant, TQM has lost favour. Even where some positive results become apparent in a short space of time, they may not always seem to be major. How though can you tell if there have been benefits and if they are significant or not?

If after adopting TQM an organisation is still in business and the results are slightly up on the previous year, is this something to be excited by? Maybe the shareholders won't see this as a triumph. But it may well be. If the organisation had not begun its quality revolution, perhaps the results would have been much worse.

Jan Carlzon

Sometimes, just a change in attitude and recognition of key problem areas can be sufficient to make a big difference. For example, when Jan Carlzon took over SAS, the airline was about to lose $20 US million. Despite this he found that Scandinavian Airlines (SAS) was a very efficient organisation. They knew their business, it was transporting goods and people by air, and they did this with clinical efficiency. They had sufficient resources and well-trained staff and those 10 million passengers were carried each year. Carlzon then established that for each passenger there were five occasions when the passenger came into contact with front line employees and that this contact lasted on average 15 seconds. He called these contact times moments of truth:

> *Last year 10 million customers came into contact with approximately five SAS employees, and this contact lasted on average of 15 seconds each time. Thus, 'SAS' is created in the minds of the customers 50 million times a year, 15 seconds at a time. These 50 million 'moments of truth' are the moments that ultimately determine whether SAS will succeed or fail as a company. They are moments when we must prove to our customers that SAS is their best alternative.*
>
> Carlzon (1989)

By establishing moments of truth and by converting the staff to his way of thinking, and by taking some positive actions within 12 months he was able to turn a $20 million loss into a $40 million profit.

But this example is an exception: few turnarounds are this dramatic, and benefits accrue over longer terms. The philosophy of TQM is to look for a continuous improvement, not major breakthroughs, any major breakthrough will be a bonus. No organisation can ever say that TQM has been achieved – the quest for improvement is never ending.

ISO 9000

In a discussion on the subject of quality it would be wrong to ignore the effect that the ISO 9000 series has had on quality. The ISO 9000 series (accrediting criteria revised 2000/2001) and the more recent 14000 environmental series have been developed over a long period of time. The origins can be traced back to military requirements, for example NATO in the late 1940s developed specifications and methods of production to ensure compatibility between NATO forces in weapons and weapons systems. In Britain the predecessor to

ISO 9000 was the British standard BS 5750 which was introduced in 1979 to set standard specifications for military suppliers.

ISO 9000 certification means that an organisation constantly meets rigorous standards, which are well documented, of management of quality of product and services. To retain certification the organisation is audited annually by an outside accredited body. ISO 9000 on the letter head of an organisation demonstrates to themselves, to their customers and to other interested bodies that it has an effect quality assurance system in place.

TQM means more than just the basics as outlined in or ISO 9000, indeed ISO 9000 could be seen as running contrary to the philosophy of TQM.

What does ISO 9000 achieve?

ISO 9000 primarily exists to give the customer confidence that the product or service being provided will meet certain specified standards of performance and that the product or service will always be consistent with those standards. Indeed some customers will insist that suppliers are ISO accredited.

There are also internal benefits for organisations that seek ISO 9000 accreditation.

Firstly, by adopting ISO 9000, the methodology of the ISO system will show an organisation how to go about establishing and documenting a quality improvement system. To achieve accreditation, an organisation has to prove that every step of the process is documented and that the specifications and check procedures shown in the documentation are always complied with. The recording and documenting of each step is a long and tedious job. Perhaps the most difficult stage is agreeing on what exactly the standard procedure is.

ISO tends to be driven from the top down and relies on documentation, checks and tests to achieve a standard, somewhat bland, level of quality assurance. TQM on the other hand, once established, relies on bottom-up initiatives to keep the impetus of continual improvement. However, as the Deming method of TQM does advocate a stable system from which to advance improvements, the adoption of the ISO 9000 approach will mean that there

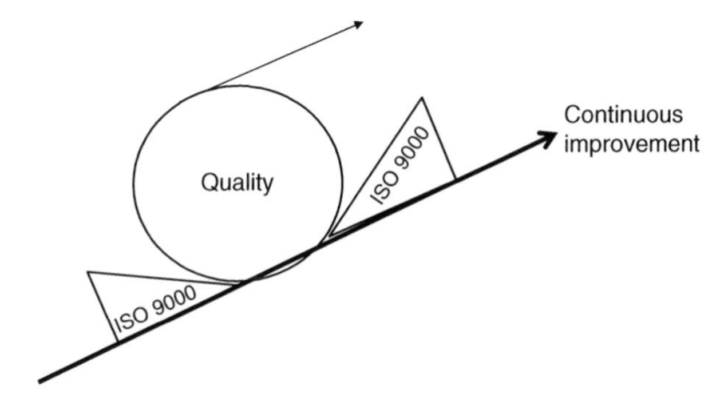

Figure 2.2 The wedge (© Ron Basu).

will be a standard and stable system. To this extent, ISO 9000 will prove a useful base for any organisation from which to launch TQM.

As shown in Figure 2.2 ISO 9000 can be depicted as the wedge that prevents quality slipping backwards, but the danger is it can also be the wedge that impedes progress.

Notwithstanding the benefits of obtaining a standard stable system through ISO procedures, it must be queried why a true quality company would need ISO 9000. If the customer or potential customer is NOT insisting in ISO accreditation, then the time and effort (and the effort expended will be a non-recoverable cost) make the value of ISO to an organisation highly questionable.

Gaining ISO 9000 accreditation is a long and expensive business. Internally it requires much time and effort, and most organisations underestimate the time and effort involved. Generally, recording the systems alone will require the full-time efforts of at least one person.

Example

One small print shop employing 20 people, and with one main customer, was sold on the idea of ISO accreditation by a consultant. They were advised that the process of obtaining accreditation would take 9 months. The actual time taken was 2 years and 3 months. The main customer had not asked for accreditation, but the difficulties experienced by the print shop in getting accredited led the customer to query the efficiency of the organisation and the account was almost lost. What of the expensive consultant? Well, they took their fee and rode off into the sunset.

Internal costs of obtaining accreditation are expensive – more expensive than most organisations are prepared to admit. Total internal costs will not be known unless everyone involved in setting up the systems records and costs the time spent, and this is seldom done. The external costs can be equally expensive. It is not mandatory to hire an external consultant, but there are advantages for doing so. Consultants are not cheap and quotes from at least three consultants should be sought. Briefing the consultants will force an organisation to do some preparatory work which if properly approached should help in clarifying the overall purpose and give some indication of the effort that will be involved. Once the consultant is employed, it will be the organisation hiring the consultant that will do the work. Consultants point the way. They give guidelines and hold meetings; they will help with the planning, but don't expect them to get their hands dirty. They won't actually do any work; the organisation seeking accreditation does the work!

Throughout the ISO 9000 series, reference is made to documentation. To meet the ISO requirements, it is not necessary to have hard copies of quality plans, quality manuals and procedures. Indeed, when people have a computer terminal at hand, they are more likely to search the computer rather than leaf through large manuals. Also, with a computer system, it is easier to update the records with the latest procedures and to ensure that the user acknowledges receipt of change when using the system. In this way the system can be kept almost instantly updated, and staff can be encouraged to make improvement suggestions.

The other important aspect of ISO is audits. The external audit requirements of the ISO 9000/14000 series are more towards compliance checks after an activity has started or been completed. This type of check confirms that procedures are being kept to, or that an outcome complies with the standard. Where mistakes are found, they are retrospective. They will highlight where errors have occurred and thus indicate the need for corrective action for the future, but they don't stop the error happening in the first place. To be effective, where internal audits are in place, the internal quality auditor should be trained in audit procedures and the purpose of auditing. Auditors should be there to help and guide, not to trap and catch. If the audit is preventative, that is before the event rather than after, so much the better.

To summarise this discussion of ISO 9000, with TQM the aim is continuous improvement and with the continuing impetus for quality improvement being driven from bottom up. ISO 9000 will not necessarily achieve this. At best, ISO can be seen as a step on the way to TQM. At worst, it might actually inhibit TQM, as it relies on the setting of top-down standards and controls and might deter staff from suggesting changes. A true TQM organisation does not need ISO but, if ISO is insisted on by a customer, it can be made to fit into the overall TQM plan.

Further details of ISO 9000 as a Quality Management System and technique for operational excellence can be found in Chapter 10.

Kaizen

The Japanese have a word for continuous improvement: it is 'Kaizen'. The word is derived from a philosophy of gradual day-by-day betterment of life and spiritual enlightenment. Kaizen has been adopted by Japanese business to denote gradual unending improvement for the organisation. The philosophy is the doing of little things better to achieve a long-term objective. Kaizen is 'the single most important concept in Japanese management – the key to Japanese competitive successes' (Imai, 1986).

Kaizen moves the organisation's focus away from the bottom line, and the fitful starts and stops that come from major changes, towards a continuous improvement of service. Japanese firms have, for many years, taken quality for granted. Kaizen is now so deeply ingrained that people do not even realise that they are thinking Kaizen. The philosophy is that not one day should go by without some kind of improvement being made somewhere in the company. The far reaching nature of Kaizen can now be seen in Japanese government and social programmes.

All this means trust. The managers have to stop being bosses and trust the staff; the staff must believe in the managers. This may require a major paradigm change for some people. The end goal is to gain a competitive edge by reducing costs and by improving the quality of the service. To determine the level of quality to aim for, it is first necessary to find out what the customer wants and to be very mindful of what the competition is doing.

The daily aim should be accepted as being 'Kaizen' that is, some improvement somewhere in the business.

Quality Circles

In the 1960s, Juran said 'The Quality Circle movement is a tremendous one which no other country seems to be able to imitate. Through the development of this movement, Japan will be swept to world leadership in quality' (Juran, 1988). Certainly, Japan did make a rapid advance in quality standards from the 1960s onwards and Quality Circles were part of this advance. But Quality Circles were only one part of the Japanese quality revolution.

Quality Circles have been tried in the United States and Europe often with poor results. From first hand experience of Quality Circles in Australasia, in the United Kingdom (Nevan), Europe, South America, Africa, Asia and India (Ron) we believe that Quality Circles will work if the following rules are applied:

1. The circle should only consist of volunteers.
2. The members of the circle should all be from different functional areas.
3. The problem to be studied should be chosen by the team, and not imposed by management. Problems looked at by the circle may not always be directly related to quality or, initially, be seen as important by management.
4. Management must whole-heartedly support the circle, even where initially decisions and recommendations made by the circle are of an apparently trivial nature or could cost the company money (such as a recommendation for monogrammed overalls).
5. The members of the circle will need to be trained in working as a team (group dynamics) problem solving techniques, and in how to present reports. The basic Method Study approach of asking Why (What, Where, When, Who and How) is a standard Quality Circle approach to problem solving and members will need to be taught how to apply this structured approach to solving problems.
6. The leader of the circle and the internal management of the circle should be decided by the members.
7. Management should provide a middle manager as mentor to the circle. The mentors role is to assist when requested and generally to provide support. The mentor does not manage the circle.

In Japan, the Quality Circle traditionally meets in its own time rather than during normal working hours. Not only do circles concern themselves with quality improvement but they also become a social group engaged in sporting and social activities. It is not expected in a European country that a Quality Circle would meet in the members' own time. Few workers are that committed to an organisation. However, there is no reason why, once the Quality Circle is up and running, management could not support and encourage social events for a circle, perhaps in recognition of an achievement.

Quality Project Teams

A problem experienced in the United Kingdom with Quality Circles was the blurring of Circles and Quality project teams. The project team approach was top down, that is management selected a hard quality problem and designated staff to be members of the team. The top-down, conscription approach might appear to be more focused than the Quality Circle approach, but the fundamental benefits of a voluntary team approach are lost. With the pure, bottom-up Quality Circle approach and the members are volunteers and the circles consist of people who work well together and who want to contribute to the success of the organisation.

Ishikawa (Fishbone technique) or Cause and Effect

The Ishikawa diagram named after its inventor Kaoru Ishikawa (1979, 1985) or the Cause and Effect Diagram is designed for group work. It is a useful method of identifying causes and provides a good reference point for brainstorming (brainstorming is discussed in a separate paragraph later in this chapter).

The usual approach is for the group to agree on a problem or *EFFECT*. Then a diagram is drawn consisting of a 'back bone' and four (sometimes more) fishbones are shown to identify likely *CAUSES*. Common starting points are *people, equipment, method, material.*

The following eight Causes cover most situations:

- Money (funding)
- Method
- Machines (equipment)
- Material
- Marketing
- Measurements
- Management and mystery (lack of communication, secret agendas, etc.)
- Maxims (rules and regulations)

Chapter 5 also includes the Cause and Effect Diagram as a tool at the 'measure' stage of a Six Sigma programme.

Compatibility with FIT SIGMA™

All of the foregoing is compatible with FIT SIGMA™ (Basu and Wright, 2003). Some of it such as ISO 9000/14000 is not necessary, but where it exists in an organisation it is not wasted and would provide a good foundation to move up to FIT SIGMA™. FIT SIGMA™ is both a philosophy and an improvement process. The underlying philosophy is that of a total business-focused approach underpinned by continuous reviews and a knowledge based culture to sustain a high level of performance. In order to implement the FIT SIGMA™ philosophy, a systematic process is recommended. This process is not a set of new or unknown tools; in fact, these tools and culture have been proven to yield excellent results in earlier waves of quality management. The differentiation of FIT SIGMA™ is the process of combining and retaining successes. Its strength is that it is not a rigid programme in search of problems, but an adaptable solution fit for any specific organisation.

FIT SIGMA™ is a new and exciting approach of harnessing and sustaining gains from previous initiatives to secure operational excellence. It is a quality process beyond Six Sigma, which combines three main principles:

1. Fitness for purpose
2. Fitness for sustainability
3. SIGMA for integration

Its main objective is to provide sustainability and company wide integration in an appropriate way for the company to continue to improve.

FIT SIGMA™ has four additional features that Lean Sigma does not display:

1. Initial review at the start to identify gaps and fitness for purpose and formal reviews at regular intervals.
2. Periodic self-assessment with a structured and agreed checklist.
3. A continuous learning and knowledge management programme.
4. Making the initiative apply across the whole business and refocusing away from variation and on to a seamless organisation.

New trends

In addition to FIT SIGMA™ there are hybrids of Six Sigma in their applications. Almost following the publication of FIT SIGMA™ by Basu and Wright (2003) Motorola University (2004) recognised that original Six Sigma started to lose its relevance in the 1990s, as it was perceived to be too complex, and only effective for the manufacturing and engineering divisions of an organisation, and introduced the concept of New Six Sigma. New Six Sigma is an improved version of Six Sigma that incorporates the lessons learned in the 1980s and 1990s and is seen as an overall business improvement method. This concept recommends five principles, Align, Target, Mobilise, Accelerate and Govern, and proposes four key goal areas:

1. Process improvement
2. Financial gains
3. Customer satisfaction
4. Innovation and growth

A performance excellence audit is used to determine any gaps in the Six Sigma implementation and determine a baseline. This then leads to an assessment of the current reality and the desired improvements needed. These are assigned to cross-functional teams. With clear definition and visible management support the teams of people are then trained as Black or Green Belts as required. The implementation is focused around weekly project reviews and monthly steering team reviews. On a close examination there many similarities between FIT SIGMA™ and New Sigma approaches.

Motorola (http://mu.motorola.com/faqs.shtml) also announced the revival of Six Sigma with Digital Six Sigma (DSS) by leveraging IT to improve training, processes, tools and tracking. The foundation for DSS is the Business Process Management System (BPMS) which provides a process platform on which process modelling and Six Sigma tools can be integrated.

DSS is reported to be a revitalisation of Six Sigma methodology through the following improvements:

- Leveraging new digital tools to drive project success.
- Digitising business processes to 'permanently' enforce optimal process compliance.
- Tracking vital processes through digital databases.
- 'Permanently' and proactively eliminating sources of variation that cause defects.

Marash (2004) proposes a concoction of well known philosophies and approaches of quality programmes and tools such as Kaizen, TQM, ISO 9000, Excellence Models, Six Sigma and Lean and names it 'Fusion Management'. It is good to have all methods in one place but the selection of an appropriate solution is likely to be challenging to organisations and may open the door for consultants. The suggested eight steps approach of 'Fusion Management' appears to be similar to the implementation steps of FIT SIGMA™ (Basu and Wright, 2003).

Plaster and Alderman (2006) suggested Customer Value Creation (CVC) with a Six Sigma approach. It focuses on creating and exchanging value with its customers and uses an analytical process based approach to aid growth. This system uses the Six Sigma rigour and process based tools and combined with an outside-in customer focus to help drive decisions for profitable growth. The system advocates an extra layer 'the platinum belt' graded above Black Belts and trained in the science of customer service, marketing, sales and business. The tools for CVC are also similar to Six Sigma tools although some of those are renamed to emphasise focus on customer value.

3

The scope of tools and techniques

It must be considered that there is nothing more difficult to carry out, no more doubtful to success, no more dangerous to handle than to initiate a new order of things.

— Machiavelli (1513)

Introduction

Tools and techniques are essential ingredients of a process. These are instrumental to the success of a quality programme. Similarly many companies have used tools and techniques without giving sufficient thought and have then experienced barriers to progress (McQuater et al., 1994). Tools and techniques can be a double-edged sword – they are effective in the right hands and can be dangerous in the wrong ones. In this chapter, we aim to address three aspects:

1. The drivers for tools and techniques
2. The problems of using tools and techniques
3. The critical success factors

Tools and techniques are not a panacea for quality problems, but rather can be seen as a means of solving them.

The drivers for tools and techniques

It is clear that an organised approach of continuous improvement will require the use of a selection of tools and techniques for any effective problem solving process. There are a number of good reasons for this including:

- They help to initiate the process
- They pinpoint the problem
- They offer a basis for systematic analysis leading to a solution
- Employees using them feel involved

- They enhance teamwork through problem solving
- They provide an effective medium of communication at all levels
- They form a single set of methodology
- They facilitate a mindset of quality culture.

With the continuous growth of outsourcing and collaborative partnership between suppliers and customers, some tools and techniques offer a common platform for supply level agreements. A customer may insist upon the use of a specific technique as part of an agreement with its supplier. For example, automotive component suppliers have developed a learning experience to apply FMEA (Failure Mode and Effects Analysis) to satisfy the customers that the technique is applied in an effective way. For this manner, both the supplier and customer share the improvement programme and enhance the mutual competitive advantage.

The following three key factors should be considered carefully when selecting tools and techniques for a quality initiative:

1. *Rigour in purpose* – The tools or technique selected must be meeting the main purpose or reason for its application. No single tool is more important in isolation, but could be most significant for a specific application. The approach must not be 'a solution in search problems'.
2. *Rigour in training* – It is imperative that all users of a tool or technique are trained to a level of competence so that they feel comfortable to apply it effectively. It is like giving someone the best golf club and expecting him to win automatically a grand tournament. It just isn't possible to become a good player without proper training and practice – otherwise, every one of us could be a Tiger Woods or a Jack Nicklaus.
3. *Rigour in application* – After the appropriate selection of a tool or technique followed by adequate training, its success will be determined by the results of its application. The key criteria are: has it solved the problem or has it improved the process? There are instances when a company created a high expectation by selling the virtues of one specific technique. A single tool or technique on its own will produce results in a limited area. It is the cumulative effect of a number of appropriate tools and techniques that would create sustainable benefits for the whole organisation: Dale et al. (1993) developed a table which would assist in identifying the specific application of tools and techniques (see Table 3.1).

The problems of using tools and techniques

There are many examples where difficulties were encountered during the application of complex techniques such as the Design of Experiments (DOE), Failure Mode and Effects Analysis (FMEA) and Quality Function Deployment (QFD). However there are also more frequent instances when the use and application of all types of tools and techniques were not addressed to the specific

Table 3.1 Application of tools and techniques

Application	Tools and techniques
1. Checking	Checklists, control plans
2. Data collection/presentation	Check sheets, bar charts, tally charts, histograms, graphs
3. Setting priorities/planning	Pareto analysis, arrow diagram, quality costs
4. Structuring ideas	Affinity diagrams, systematic diagrams, brainstorming
5. Performance/capability measurement/assessment	Statistical process control, departmental purpose analysis
6. Understanding/analysing problems/process	Flow chart, Cause and Effect Diagrams, PDPC
7. Identifying relationship	Scatter diagrams/regression/correlation/ matrix diagrams
8. Identifying control parameters	Design of experiment
9. Monitoring and maintaining control	Mistake proofing, FMEA, matrix data analysis
10. Interface between customer needs and product features	Quality function deployment

Source: Dale et al. (1993).
PDPC: Process Decision Programme Chart.

requirements of companies and their people using them. There are common problems in the use of all tools and techniques. Dale et al. (1993) identified five areas of difficulties, viz., resources, management commitment, detection based mentality, knowledge and understanding and resistance to change.

Our research (Basu and Wright, 2003) has pinpointed clearly four key factors leading to problems for the effective application of tools and techniques. These are:

1. Inadequate training
2. Management commitment of resources
3. Employee mindset
4. Poor application of tools and techniques.

Let us address these areas in a little more detail.

1. Inadequate training

The proper training of employees for the use of tools and techniques depends on three key factors:

1. Technical facts transfer
2. Organisational culture and education
3. Qualified trainers.

Most of the tools and techniques are conceptually simple but detail rich. It is important to separate the technical aspect of a tool or technique and transfer the facts in the context of the known environment of the organisation. It is very helpful to demystify a concept or definition by a worked out example based upon familiar data from the company.

The users of the tools are usually derived from a different level of the organisation and therefore the learning process must be geared to the specific capability and culture of the organisation.

The initial training is usually carried out by specialists or external consultants. The employees are likely to find the training more sustainable when it is carried out by their line managers or colleagues who have been trained as trainers. Further more, if the training depends on external consultants then the learning ceases with the departure of these consultants.

2. Management commitment of resources

Members of Senior Management are often unsure what they may expect from tools and techniques. There is also the existence of rivalry between different departments and middle managers fail to see the benefits of a change beyond their own turf. Consequently, adequate resources are not always made available to support the training and improvement activities arising from the application of tools and techniques.

In order to have any success with the use of tools and techniques it is essential to ensure total commitment and leadership from the top management. This form of support is more significant for basic tools for company-wide applications (e.g. Cause and Effect Diagram, Pareto analysis, Flow Diagram). For more advanced and specialist tools and techniques (e.g. DOE, FMEA, QFD), departmental involvement is even more important.

3. Employee mindset

In congruence with the commitment of senior management, another important hurdle encountered while implementing tools and techniques is the mindset of all the employees of a company. As part of a company-wide quality programme, a group of people are brought together who may have never worked together before to achieve a novel task. In spite of the full support of top management, failure to convince the project stakeholders may cause your project to fail. In the case of tools and techniques, the main stakeholders are employees. Managing people is a complex area encompassing such considerations as human motivation, attitude and culture. There are an abundance of published works available for the reader who is interested in pursuing this area in greater depth. For our specific purposes regarding the problems of using tools and techniques, three models or approaches are most relevant.

Tuckman et al. (2001) outline that members of the team have to develop a set of common values or norms before they can work together effectively as a group. As shown in Figure 3.1, project teams typically go through five stages, which can be dubbed: forming, storming, norming, performing and mourning.

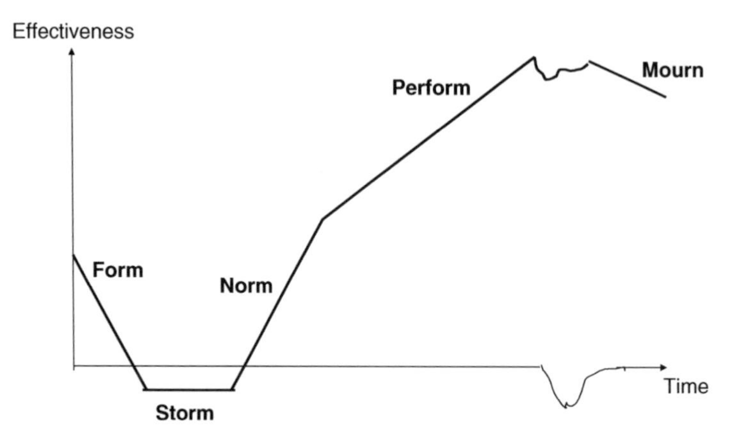

Figure 3.1 The Tuckman model of team formation (© Ron Basu).

At *forming*, the team members come together with a sense of high anticipation and motivation. As the team begins to work together, differences occur during the '*storming*' stage and the team's level of performance drops. Gradually the team develops a sense of group identity and a set of values or *norms*. The team members start to work together and their motivation and effectiveness begin to increase and thus group *performance* reaches a plateau. Finally towards the end of the project where the future of the team is uncertain, the *mourning* stage sets in.

The second model is based upon McElroy and Mills (2000). The two attributes of employees, commitment and knowledge, are mapped on to a chart, as shown in Figure 3.2, to group the knowledge base across employees into four quadrants. To populate this chart certain assumptions are made

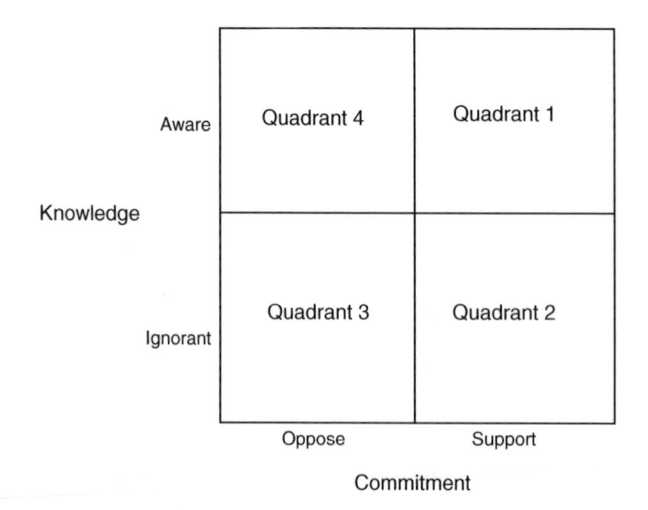

Figure 3.2 Knowledge–commitment matrix (© Ron Basu).

regarding the knowledge base and attitudes of employees. These assumptions could be tested by a voluntary survey.

- *Quadrant 1*: Support/aware – These supporters are the key players for tools and techniques.
- *Quadrant 2*: Support/ignorant – This support is vulnerable and should be reinforced by training.
- *Quadrant 3*: Oppose/ignorant – This is a key target area to achieve commitment by a combination of culture and facts transfer.
- *Quadrant 4*: Oppose/aware – These will be the most difficult group of employees to convert. The only way to move them to support the project could be to negotiate a training role for them.

In the third model, Wallace (1990) suggests that the distribution of three groups of employees (Naysayers, the Silent Majority and Enthusiasts) can be transformed by a properly administered education programme for employees. This change is illustrated qualitatively in Figures 3.3 and 3.4.

Even after education, there will be a small group of people (the 'naysayers') who do not believe that tools and techniques will actually work.

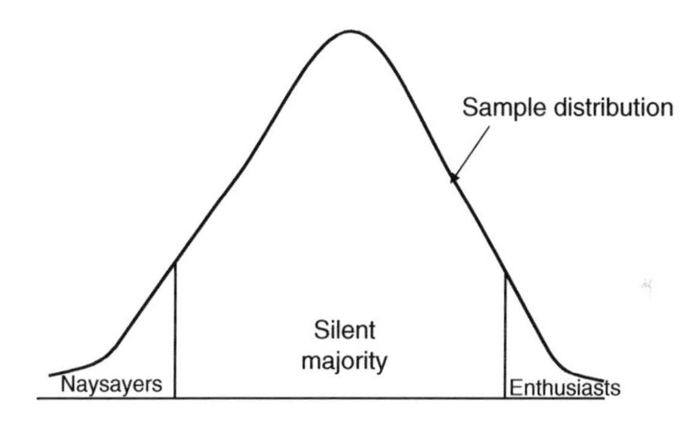

Figure 3.3 Commitment before education (© Ron Basu).

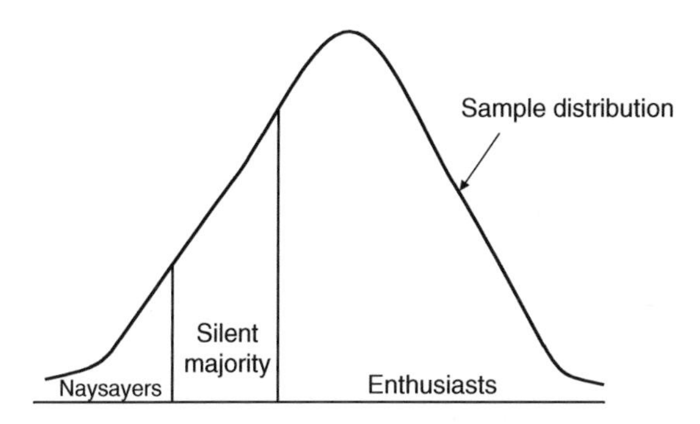

Figure 3.4 Commitment after education (© Ron Basu).

4. Poor application of tools and techniques

The inappropriate application of tools and techniques introduces the additional problems of using them to achieve and sustain results. This is often a more difficult area to correct when the user has applied a tool which is specific for a departmental target. For example, a packing line supervisor is striving to maximise the OEE (Overall Equipment Effectiveness) of a particular soap wrapping machine, ACMA 791. The supervisor's aim will be to maximise output and minimise the loss time of the machine. As a consequence, he or she is likely to produce more than the scheduled quantity. This creates an obstacle to achieving the scheduling performance. The appropriate application of tools and techniques aims to improve and sustain the business performance as a whole rather than conflicting departmental efficiency. An effective way to overcome these 'turf management' issues is to ensure that training includes the understanding of the holistic approach of using tools and techniques for the business rather than the department.

The typical signs of the poor application of tools and techniques are:

- Department Managers are using tools and techniques as a routine to publicise departmental performance rather than as a basis to identify deficiencies for improvement.
- Dependence on computer softwares rather than the rigour of data collection, correct input and interpretation of results.
- Using tools and techniques to find excuses for not making changes.
- Knowledge is restricted to specialists and not shared by relevant employees.

The critical success factors

The factors influencing the success of tools and techniques are naturally contributed by the 'drivers for tools and techniques'. However, in spite of the apparent paradox, the 'problems' of using tools and techniques also contribute to the success by offering restraining forces. The understanding and anticipation of problems at an early stage ensure the avoidance of pitfalls during a later aspect of the application. The restraining forces provide the basis for both quality assurance and quality control to the drivers of tools and techniques leading to critical success factors as shown in Figure 3.5.

There are a number of resultant success factors in the effective use of tools and techniques and these are mostly in common with success factors for a company-wide change programme. The critical success factors are:

1. Top management commitment
2. Availability of resources
3. Well-designed education and training programme
4. Rigorous project management approach.

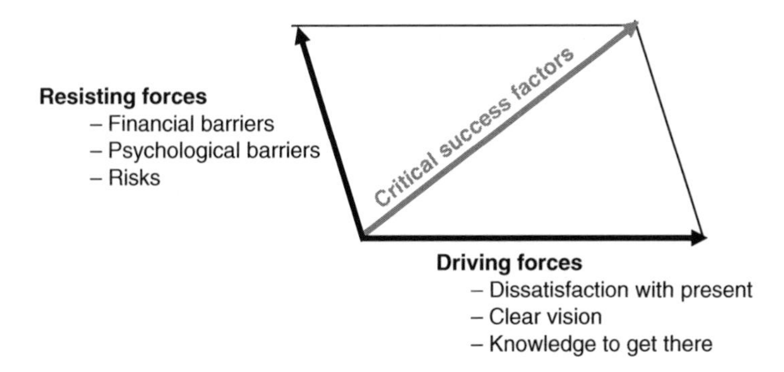

Figure 3.5 Forces of change management (© Ron Basu).

1. Top management commitment

Various case studies and research reports have concluded over and over again that without the total commitment of senior management, a company-wide project or change programme can never succeed. The essential requirement is that the top management must be convinced that there is a business case for the use and application of tools and techniques. The secondary requirement is that they can be embedded in the corporate programmes. Critical to the successful introduction of tools and techniques is a champion, a senior manager whose position, enthusiasm and knowledge can be used to good effect.

2. Availability of resources

It is paramount that, once a decision has been made by the management to implement tools and techniques as part of a company-wide quality programme, the management ensures that all relevant resources for the initiative are available and used effectively. Resources in this context relate to people, facilities, systems, cost and money. The availability of one's own employees is most critical, because the demands of the 'day job' to meet customer-dominated business requirements often introduce a conflict between the project and the operation. The 'people resource' is required at two levels: they are necessary as project team members or trainers as well as other employees fulfilling the role of users. A good practice of resource utilisation is to train a small group of carefully selected people in tools and techniques and then use them to transfer their knowledge to other employees of the company. This approach has the advantage of both reducing the cost of consultants and external trainers, and customising the training to the needs and culture of the company.

3. Well-designed education and training programme

Education is fundamental to making things happen. Education comprises both the 'facts transfer' related to tools and techniques and the 'behaviour change' necessary for employees to do their jobs differently. Education lies in teaching employees how to use the tools and getting them to accept that they can work with these tools as a team.

Training is the process of making smaller group of people become proficient in the more complex techniques, such as QFD, FMEA and DOE.

Many Total Quality Management (TQM) education and training programmes have failed because of their emphasis on cultural issues and were nicknamed the 'country club approach.' The rigorous training deployment plan of Six Sigma is a good model for a well-designed programme. Employees are educated and trained following defined modules and then assessed to gain 'Green Belt' or 'Black Belt' status. Green Belts are educated in tools while Black Belts are additionally trained in the application of techniques.

4. Rigorous project management approach

The success of the tools and techniques are further assured when these are used as part of a well managed project or programme. The rigours of project management in the context of this discussion must include:

- A well-defined project brief
- Project appraisal before any major commitment
- Project organisation
- Regular monitoring of time, cost and benefits
- Continuous feedback, learning and sharing of good practice.

Common failures are the over-run in time and budget as well as incomplete achievement of the original objectives of the projects. Projects generally have five basic elements: scope, time, cost, quality and risk. With the use and application of tools and techniques, we take a whole systems approach to each of the five basic elements and the related issues listed above.

Summary of Part 1

The first chapter covers the concept of quality and quality management. Quality has three dimensions – product quality, process quality and organisation quality. The cost of quality can be measured in terms of prevention cost, appraisal cost, internal failure cost and external failure cost. Quality management has evolved over the years from inspection to control to assurance to TQM. The new waves of Six Sigma, Lean Processes and FIT SIGMA™ are embedded in the holistic programmes of Operational Excellence. We need both tools and techniques in quality management and Operational Excellence programmes.

Chapter 2 has covered the history of quality and considered the contributions of quality gurus since the 1960s, and considered some approaches and basic tools used in various quality systems.

Our belief is that quality is not a new or separate discipline, that quality pervades all management actions. Our philosophy is that quality is too important to be left to the managers and that quality is everybody's concern, not only in the organisation, but also the concern of customers and suppliers and any other stakeholder.

Quality has two main aspects: it can be measured from the customer's perspective–customer satisfaction, and it can be viewed from the perspective of efficient use of resources. These two seemingly separate objectives are in fact inseparable when quality is considered. An organisation that wishes to compete in the global market must be efficient and provide a high level of customer satisfaction. No organisation will be able to afford to provide world class service unless its use of resources is efficient and non-value adding activities have been minimised, and no organisation cannot afford to be world class.

In this chapter we also briefly considered the part ISO 9000/14000 has in a total quality approach. Specific techniques such as quality circles and cause and effect analysis were also introduced. Finally it was shown that elements of all the quality initiatives over the last 40 years are compatible with FIT SIGMA™. FIT SIGMA™ is the new wave that enables an organisation to maintain operational fitness.

In Chapter 3 we have reviewed both the positive drivers and the problems conducive to restraining forces for the use and application of tools and techniques. We have concluded that success factors are derived from the resultant vector of the mutually contradicting driving forces and restraining forces. The critical success factors are: top management commitment, availability of resources, well-designed education and training programmes and a rigorous project management approach.

Part 1: Questions and exercises

1. Explain the distinctive features of the three dimensions of quality – product, process and organisation. Discuss and distinguish between the dimensions of quality as presented by Gravin and Parasuraman.
2. Identify from the customer's viewpoint those dimensions of quality which could be important in the following operations:
   (a) QE2 cruise
   (b) Medical centre
   (c) Supermarket service
   (d) Sale of a Rolex Watch
   (e) Package holiday
   (f) Ford Motor Company
   (g) Norwich Union Insurance Company
   (h) Advertising of a Nokia Mobile Phone
3. What methods or systems does your organisation have in place for the quality management of the input, process and output of operations?
   (a) How well are the standards known to all members of the organisation? Are the suppliers regarded as part of the quality management process?
4. How would you distinguish between inspection, quality control, quality assurance and total quality management?
   (a) What are the appropriate areas or stages of application for each scheme? What does 'total' mean in TQM?
5. Describe a model for Total Quality Management (TQM) and compare it with that for Business Process Re-engineering (BPR).
   (a) What are the 'five roles' in a BPR programme? Write the job description of a so-called Czar.
6. What are the features and philosophy common to both TQM and Six Sigma?
   (a) Explain the new features, if any, in a Six Sigma programme. What are the additional features in Lean Sigma?
7. What are Deming's '14 Criteria for Quality Improvement'?
   (a) Explain Deming's 'Quality Wheel'. With appropriate reference to the 'quality guru', explain the following:
      – 'Fitness for Purpose'
      – 'Quality is Free'
      – 'Moment of Truth'
      – 'Fishbone Diagram'
      – 'Single Minute Exchange of Dies'
      – 'FIT SIGMA™'
8. Explain the various elements of the 'cost of poor quality'.
   (a) Why is it that some quality-related costs, after the delivery of goods, are more significant to the supplier? List the key elements of the internal and external failure costs in your organisation.

9. Explain why tools and techniques are essential in the implementation of quality and operational excellent programmes.
   (a) How do you distinguish between a tool and a technique?
10. Describe the main driving and resisting forces in the implementation of a change programme.
   (a) What are the critical success factors related to the application of tools and techniques?

Part 2

Tools

Introduction to Chapters 4–8

In our introduction in Chapters 1 and 3, we have defined 'tools and techniques' for quality management. In the following five chapters (i.e. Chapters 4–8) we describe the useful features of tools and then we deal with techniques in Chapters 9 and 10.

A single tool is a device which has a clear role and defined application. However, numerous tools jostle for space in the training literature. Our experience has suggested that although most of these tools are basically simple, their inappropriate selection and application often ended in failure. Thus we need to present them in a structured way, and this is the key to our approach. They can be structured either in clusters such as QC7 (Quality Control 7) and M7 (Management 7), or as part of an improvement cycle, such as Deming's PDCA (Plan, Do, Check, Act) and Six Sigma's DMAIC (Define, Measure, Analyse, Improve, Control).

DMAIC (see Chapter 9 for details) provides the rigour of a proven project management life cycle. Furthermore because of its extensive use in ever-growing Six Sigma initiatives, DMAIC will probably be recognised and understood quite easily. We have therefore arranged the next chapters according to the DMAIC model as follows:

Chapter 4: Tools for definition (Define)
Chapter 5: Tools for measurement (Measure)
Chapter 6: Tools for analysis (Analyse)
Chapter 7: Tools for improvement (Improve)
Chapter 8: Tools for control (Control)

Although we present the tools under specific stages of DMAIC, it is to be noted that some tools can be used in more than one stage. For example, the Flow Process Chart can be deployed at all stages, but we have included it in Chapter 5 where it is likely to be most frequently used.

Each of the tools will be presented according to the following format:

- Definition
- Application
- Basic steps
- Worked-out examples
- Training requirements
- Final thoughts

4

Tools for definition

The greatest challenge to any thinker is starting the problem in a way that will allow a solution.

— Bertrand Russell

The project definition is arguably the most important part of the project life cycle, because it establishes the basis for the other project management subprocesses. In the context of a Six Sigma or related Operational Excellence programme, if we follow DMAIC as the basis of the life cycle then the 'Define' stage sets the baseline which allows the subsequent stages to follow a structured methodology and expected quality standards.

The tools at this 'Define' stage are primarily for data collection which influences the management of the project start and terms of reference. The important tools for data collection include:

D1: IPO Diagram
D2: SIPOC Diagram
D3: Flow Diagram
D4: CTQ Tree
D5: Project Charter

The key deliverable of the Define stage is an agreed Project Charter.

D1: IPO Diagram

Definition

An Input-Process-Output (IPO) diagram, also known as a general process diagram, provides a visual representation of a process by defining a process and demonstrating the relationships between input and output elements.

The input and output variables are known as 'factors' and 'responses', respectively.

Application

Whether we are performing a service or manufacturing a product or completing a task, an IPO diagram is very useful to define a process as an activity that transforms inputs in order to generate corresponding outputs.

An IPO diagram is very often the starting point of a Six Sigma or similar improvement project. This high level mapping of the processes is then followed by Flow Diagrams, Process Mapping and Design of Experiments (DOE) to understand fully the process and related sub-processes.

It is necessary to develop an IPO diagram to determine the factors and responses before carrying out a DOE exercise (see Chapter 9).

Basic steps

1. In building an IPO diagram we first choose a process.
2. Next we define the outputs or responses. They are also called 'quality characteristics' or critical to quality (CTQ) or *y*-variables. They are usually defined from a customer perspective.
3. We then define the input factors which will be required to make the process valuable to the customer.
4. Draw the IPO diagram with incoming arrows for inputs and outgoing arrows for outputs (See Figure 4.1).

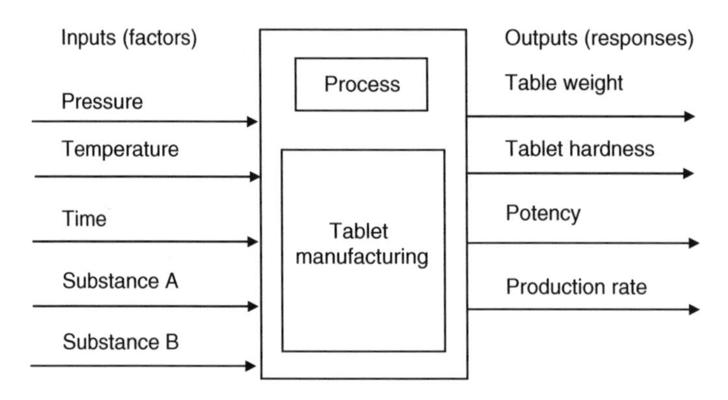

Figure 4.1 An IPO diagram.

Worked-out example

Figure 4.1 shows an example of an IPO diagram of a tablet manufacturing process in a pharmaceutical company.

Training requirements

An IPO diagram is a simple tool in principle and a 1-hour training session (including exercises) should be adequate for a project team member to develop a practical diagram.

Final thoughts

An IPO diagram is a simple tool to define a process and focus on its key variables. This is closely linked with a Cause and Effect diagram and DOE. We recommend the extensive use of IPO diagrams.

D2: SIPOC Diagram

Definition

SIPOC is a high level map of a process to view how a company goes about satisfying a particular customer requirement in the overall supply chain. SIPOC stands for:

Supplier: The person or company that provides the input to the process (e.g. raw materials, labour, machinery, information, etc.). The supplier may be both external and internal to the company.

Input: The materials, labour, machinery, information, etc. required for the process.

Process: The internal steps necessary to transform the input to output.

Output: The product (both goods and services) being delivered to the customer.

Customer: The receiver of the product. The customer could be the next step of the process or a person or organisation.

Application

A SIPOC diagram is usually applied during the data collection of a project or at the 'Define' stage of DMAIC in a Six Sigma programme. However, its impact is utilised throughout the project life cycle.

SIPOC not only shows the interrelationships of the elements in a supply chain, but also CTQ indicators such as 'delivered in 7 days'.

Basic steps

1. Select the process for the SIPOC diagram and identify CTQ parameters.
2. Determine the input requirements.
3. Identify the suppliers for each of the input elements.
4. Define the output and validate CTQ parameters.
5. Identify the customers.
6. Draw the SIPOC process diagram.
7. Retain the diagram for the rest of the improvement project.

Worked-out example

Figure 4.2 shows a SIPOC diagram for a company that leases equipment (adapted from George, 2002, p. 185).

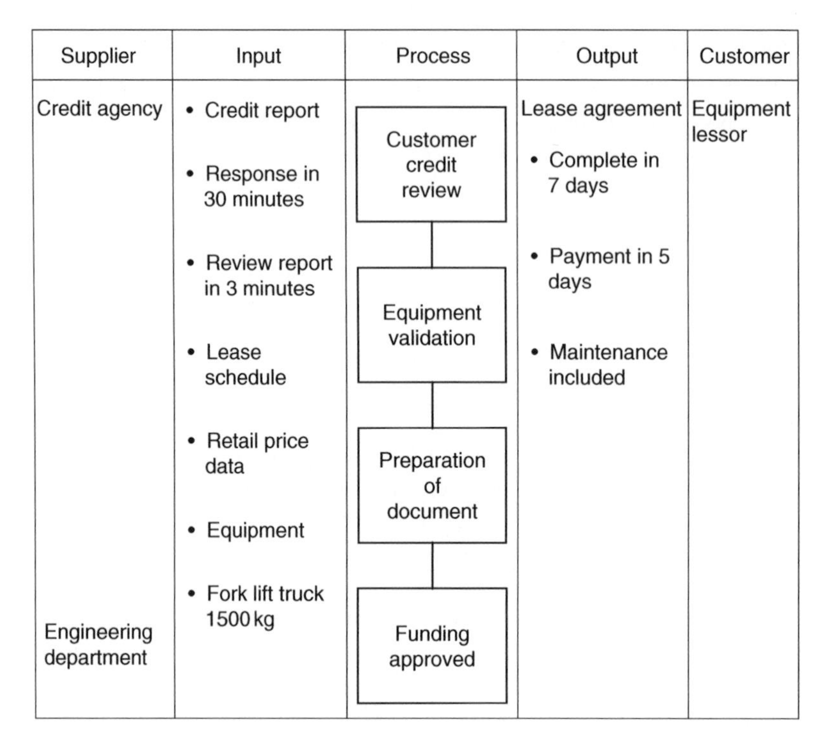

Figure 4.2 SIPOC process diagram.

Training requirements

The training of SIPOC should be combined with those for the CTQ tree and the IPO diagram as they complement one another. The combined training is expected to be carried out for both Black Belts and Green Belts in half a day.

Final thoughts

SIPOC is a useful tool to identify the customer, stakeholders and CTQs of a process before the development of a Project Charter. However it is not an essential tool and a process can be analysed at a high level by an IPO or a Flow Diagram.

D3: Flow Diagram

Definition

A Flow Diagram (also called a Flow Chart) is a visual representation of all major steps in a process. It helps a team to understand a process better by identifying the actual flow or sequence of events in a process that any product or service follows.

There are variations of a Flow Diagram depending on the details required in an application. We have included two other forms of diagrams in this family. These are the Flow Process Chart (see Chapter 5) and Process Mapping (see Chapter 6).

The type of Flow Diagram described in this section is a top level mapping of the general process flow and uses four standard symbols:

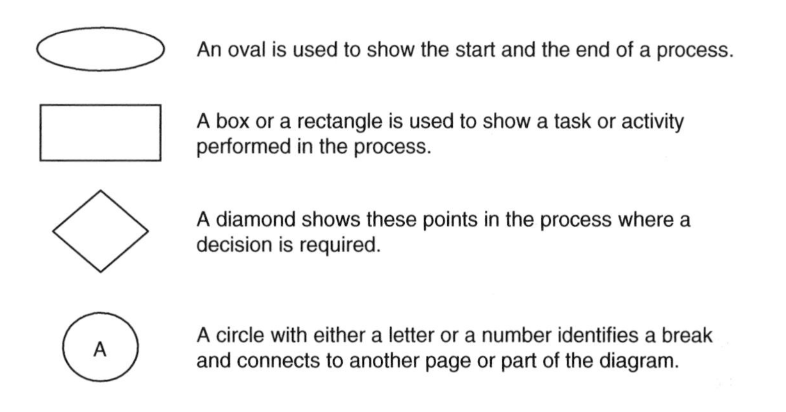

Application

A Flow Diagram can be applied to any type of process, to anything from the development of a product to the steps in making a sale or servicing a product.

It allows a team to come to a consensus regarding the steps of the process and to identify critical and problem areas for improvement.

In addition, it serves as an excellent training aid to understand the complete process.

Basic steps

1. Select the process and determine the scope or boundaries of the process:
   a. Clearly define where the process understudy starts and ends.
   b. Agree the level of detail to be shown on the Flow Diagram.
2. Brainstorm a list of major activities and determine the steps in the process.
3. Arrange the steps in the order they are carried out. Unless you are developing a new process, sequence what actually is and not what should be.
4. Draw the Flow Diagram using the appropriate symbols.
5. There are a number of good practices when charting a process including:
   a. Use Post-it™ notes so that you can move them around in a large process.
   b. For a large scale process, start by charting only the major steps or activities.
   c. Come back to develop further details for major steps if necessary.
   d. Consider Process Mapping (see Chapter 6) to apply to a larger process.

Worked-out example

Figure 4.3 shows an example of a Flow Diagram to illustrate an advertising process.

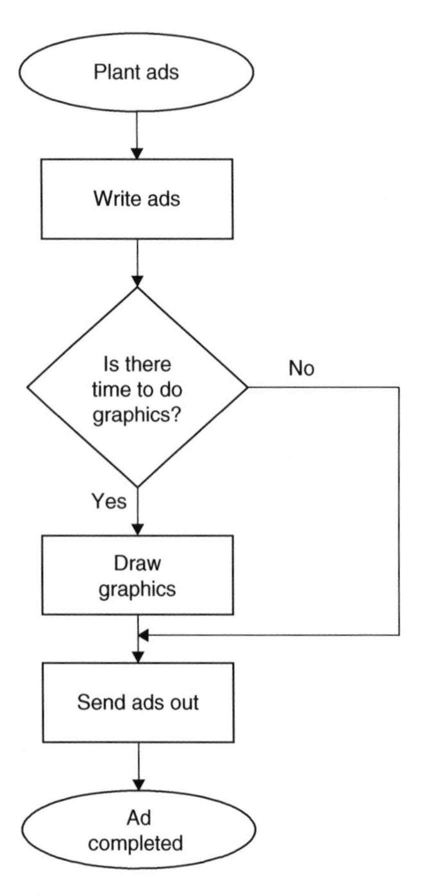

Figure 4.3 A Flow Diagram.

Training requirements

The basic principles of a Flow Diagram are very simple and can be acquired in a self-teaching process by the user. However, good experience is required to develop a Flow Diagram of a complex process. This experience can only gained by working as a team in improvement projects.

Final thoughts

We recommend the use of a Flow Diagram as a tool to map the major steps of a process. This should be useful at the start of a project to understand the process and agree on the steps of the process. For a detailed assessment or analysis of the process, a Flow Process Chart or a Process Mapping tool will be more appropriate.

D4: CTQ Tree

Definition

'CTQ' is a term widely used within the field of Six Sigma activities to describe the key output characteristics of a process. An example may be an element of a design or an attribute of a service that is critical in the eyes of the customer.

A CTQ tree helps the team to derive the more specific behavioural requirements of the customer from his general needs.

Application

A CTQ tree is a useful tool during the data collection stage (Define) of an improvement project. Once the project team has established who their customers are, the team should then move towards determining the customer needs and requirements. The need of a customer is the output of a process. Requirements are the characteristics to determine whether the customer is happy with the output delivered. These constitute what is 'CTQ' and a CTQ tree helps to identify these CTQs in a systematic way.

Basic steps

1. Identify the customer.
2. Identify customers' general needs in Level 1.
3. Identify the first set of requirements for that need in Level 2.
4. Drill down to Level 3 if necessary to identify the specific behavioural requirements of the customer.
5. Validate the requirements with the customer. The process of validation could be one-to-one interviews, surveys or focus groups depending on the CTQ.

Worked-out example

(Adapted from Eckes, 2001, p. 55)

Figure 4.4 shows a worked-out example of a CTQ tree for room service in a hotel.

Training requirements

A CTQ tree is a simple tool in principle and a 1-hour training session (including an exercise) should be sufficient for a team member to get involved in developing a practical CTQ tree. This is appropriate for both Black Belts and Green Belts.

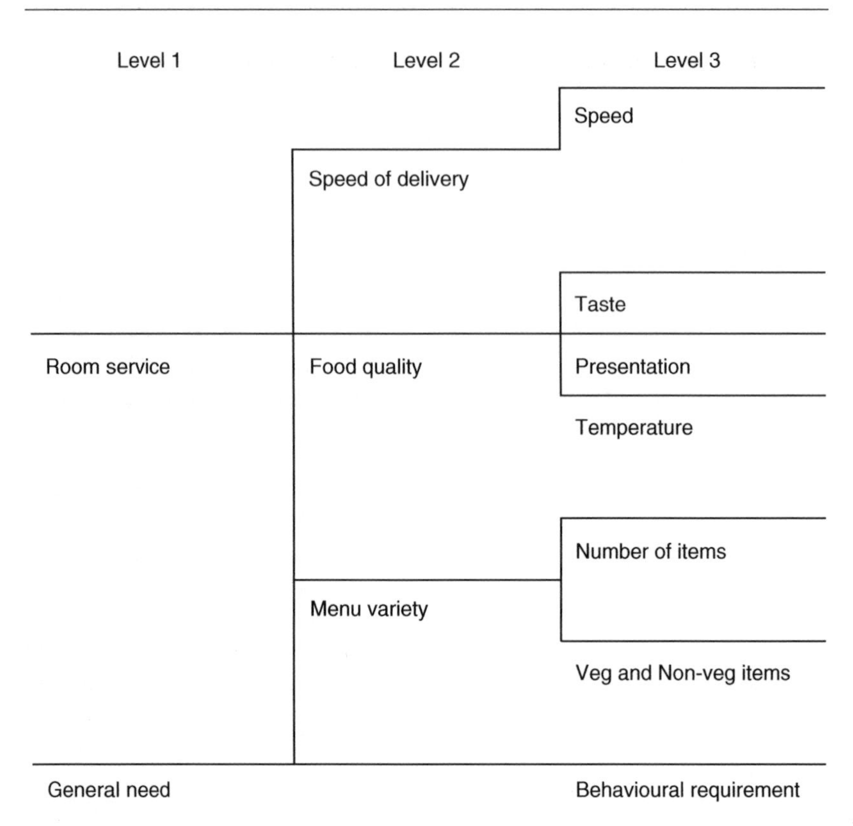

Figure 4.4 A CTQ tree.

Final thoughts

A CTQ tree is a simple but powerful tool for capturing the details of customer requirements and we recommend its use at the very early stage of a Six Sigma project.

D5: Project Charter

Definition

A Project Charter is a working document for defining the terms of reference of each Six Sigma project. The charter can make a successful project by specifying necessary resources and boundaries that will in turn ensure success.

The necessary elements of a Project Charter include:

Project Title: It is important to use a descriptive title that will allow others to quickly identify the project.

Project Type: Whether it is for quality improvement, increasing revenue or reducing cost.

Project Description: A clear description of the problem, the opportunity and the goal.

Project Purpose: Why you are carrying out this project.

Project Scope: Project dimensions, what is included and not included.

Project Objectives: Target performance improvement in measurable terms.

Project Team: Sponsor, Team Leader, and Team Members. It is important to identify Black Belts and Green Belts within the team.

Customers and CTQs: Both internal and external customers and CTQs specific to each customer.

Cost Benefits: A draft business case of savings expected and the cost required to complete the project.

Timing: Anticipated project start date and end date. Target completion dates of each phase (e.g. DMAIC) of the project are useful.

Application

A Project Charter is the formalised starting point in the Six Sigma methodology. It takes place in the Define stage of DMAIC.

We recommend that larger (e.g. over $1 million savings) projects of a Six Sigma programme which is usually led by a Black Belt should have a Project Charter. A Project Charter can also be useful in a multi-discipline, medium-sized project.

A Project Charter should be used as a working document that is updated as the project evolves. The version control of the charter is therefore very important.

Basic steps

1. Select the project by taking into account the following criteria:
   a. Not capital intensive
   b. Achievable in 6 months
   c. High probability of success
   d. Good fit with Six Sigma techniques
   e. Clearly linked to real business need
   f. Historical and current data accessible.
2. Identify the customers and their specific CTQ requirements by a SIPOC diagram.
3. Estimate an order of magnitude figures for the costs and savings for the project. (Theses estimates would be updated when more accurate data becomes available as the project advances.)
4. Obtain top management support and sponsorship.
5. Select the project team with clear leadership and ownership for delivery.
6. Develop the Project Charter following a defined template (see Figure 4.5).
7. Obtain the approval of the sponsor.
8. Review and update the charter with the necessary version control as the project progresses.

Worked-out example

Figure 4.5 shows an example of a Project Charter for a 'Safety Performance' project of DuPont Teijin Films Ltd., UK.

Project:	Safety Performance Predictor Model				
Date:	3/28/00				
Project Type:	Quality X Revenue Cost reduction				
Problem Statement:	Plant needs early warning signals so programmes can be put in place to prevent injuries/incidents				
Goal Statement:	Predict injuries/incidents prior to occurring Performance level: 1.47 DPMO Time frame: 6 months				
Project Scope and Approach:	Study safety data to determine what factors influence injuries and incidents. Use the most significant factors and develop a predictor model to serve as early warning signals for a potential deterioration in our safety performance.				
Team Members:	Terry Leadership John Debbie Leader: Vickie Sponsor: Terry Mentor: Tim				
Customers:	Leadership Terry Network Leaders				
CTQs:	Early warning signals of downward shifts in safety climate Combined data source for all site safety data To be able to identify problem areas before injuries occur				
Defect Definition:	Injuries and incidents				
Opportunities per unit:	Exposure hours				
Approximate DPMO:	1.47		Z st: 6.19		
Goal DPMO:	1.25		Z st goal: 6.19		
Stake:	Confidence Interval:	Upper:		Lower	
Capital:					
Timing:	Define	Measure	Analyse	Improve	Control
Target Completion:	04/30/00	06/30/00	07/31/00	08/31/00	09/30/00
Data Issues:	Safety Climate Indicator measurements system needs to be validated				

Figure 4.5 Six Sigma Project Charter.

Training requirements

The training of a Project Charter is included in the training programmes for both Black Belt and Green Belt attainment. The basic principles of a Project Charter are relatively simple and can be explained to team members in less than an hour. The development of a chapter with appropriate data may need a couple of days.

Final thoughts

A Project Charter is an essential working document of larger Six Sigma projects. The success of a project may often be determined by the Project Charter at the Definition stage. However for smaller projects and those of a shorter duration (e.g. less then 3 months), a Project Charter is not essential and may even create unnecessary bureaucracy.

Summary

The starting process of a project is characterised by the pressure to 'kick off' the task as soon as possible, but there is a danger of too quick a start leading to a poor foundation. The simple data collection tools as described in this chapter should help the project team to identify the key requirements of defining the project on the right basis. The use of these tools does not require any specific expertise or lengthy training but it does add significant value to develop the Project Charter.

5

Tools for measurement

When you can measure what you are speaking of, and express it in numbers, you know that on which you are discoursing. But if you cannot measure it and express it in numbers, your knowledge is of a meagre and unsatisfactory kind.

— Lord Kelvin

Introduction

It is generally accepted that when we start a journey, the most essential thing to know is where we are going. We agree that a road map showing where we want to go is important and we have covered this aspect in the section on Project Charters. Even if you have a reliable road map, you will get lost if you do not know where you are at the time of reading it! Therefore, it is vital to know the current 'as is' situation of your process, i.e. where you are now. The tools for measurement are aimed at doing just that.

The measurement stage means turning the ideas and objectives of the Project Charter into a structured appraisal process. Resources will be expanded by validating ideas and identifying further opportunities. It is during the measurement stage that the business case for the principal options of a quality improvement project is developed. Another key objective of this stage is to measure the current performance of the core business process involved in the project. The main deliverables of the 'Measure' stage are:

1. Data for the project – data collected using a data collection plan to map the current performance of the process.
2. Improvement goal for the project – specific goal or standard for improving the process performance.
3. Process capability of the project – a quantitative assessment of how well the current process meets the performance standard of the project.

The important tools for measurement should include:

M1: Check Sheets
M2: Histograms
M3: Run Charts
M4: Scatter Diagrams
M5: Cause and Effect Diagrams
M6: Pareto Charts
M7: Control Charts
M8: Flow Process Charts
M9: Process Capability Measurement

The techniques for the Measure stage include Statistical Process Control (see Chapter 9) and Benchmarking (see Chapter 10).

M1: Check Sheets

Definition

The check sheet is a simple and convenient recording method of collecting and determining the occurrence of events. These sheets or forms allow a team to systematically record and compile data from observations so that trends can be shown clearly.

Application

The check sheets are very easy to apply and are used to record non-conforming data and events including:

- The breakdown of machinery
- Non-value added activities in a process
- Mistake or defects recording in a process or a problem.

The forms are prepared in advance of recording the data by the operatives being affected by the problem. It makes patterns in the data clearer, based on facts, from a simple process that can be applied to any performance area. The supplement attribute quality control charts and histograms in a complex process provide a form of 'mistake proofing'.

Basic steps

1. Agree on the type of data to be recorded. The data could relate to the number of defects and type of defects and apply to equipment, the operator, process, department, shift, etc.
2. Decide which characteristics and items are to be checked.

3. Determine the type of check sheet to use, e.g. tabular form, defect position or tally chart.
4. Design the form to allow the data to be recorded in a flexible and meaningful way.
5. Decide who will collect data, over which period and from what sources.
6. Record the data on check sheets and analyse data.

Worked-out examples

The design of check sheets is highly flexible but they can be grouped into three main categories: tabular form, defect position and tally charts.

Figure 5.1 shows an example of a check sheet in a tabular form.

Check items	Week number							
	Day 1	Day 2	Day 3	Day 4	Day 5	Day 6	Day 7	etc.
Incorrect brand specification								
Incorrect print								
Density								
Ink smudging								
Mis-registration								

Figure 5.1 Example of check sheet in tabular form.

Figure 5.2 shows an example of a check sheet where defects positions are marked on the drawing of a product.

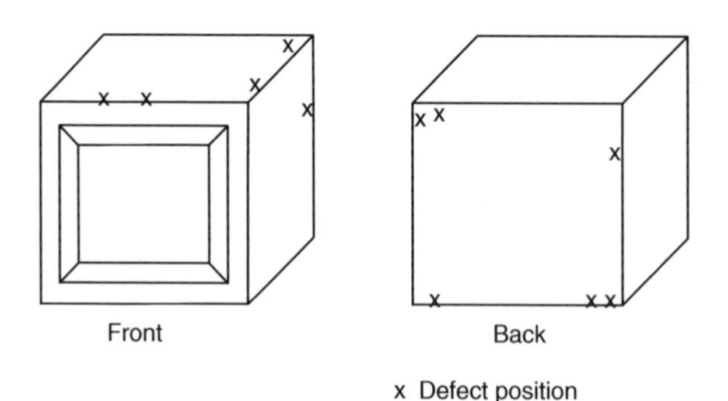

Figure 5.2 Example of a check sheet to illustrate defect positions (© Ron Basu).

Figure 5.3 shows an example of a check sheet.

Mistakes	Check sheet for: Typing February 03				
	Week 1	Week 2	Week 3	Week 4	Total
Centring	11	1111	11	~~1111~~	13
Spelling	~~1111~~ 1	1111	11	111	15
Punctuation	~~1111~~ 1111 11	~~1111~~ ~~1111~~ 1111 1	111	1111	35
Missed paragraph	11	1	1	111	7
Wrong page number	1	1	11	1	5

Figure 5.3 Example of a check sheet as a tally chart.

Training requirement

The basic principles of a check sheet are so simple that it does not require any classroom training. A member of a project usually gains experience in a check sheet by on-the-job training.

Final thoughts

Check sheets are very simple to apply and effective as a tool for data collection and team building. This is a case of low investment and high return, and therefore cannot be missed.

M2: Histograms

Definition

A histogram is a graphical representation of recorded values in a data set according to frequency of occurrence. It is a bar chart of numerical variables giving a graphical representation of how the data is distributed.

Applications

The histogram is used extensively in both statistical analysis and data presentation. A histogram displays the distribution of data and thus reveals the amount of variation within a process. There are a number of theoretical models for various shapes of distribution of which the most common one is the normal or Ganssion distribution.

There are several advantages of applying histograms in continuous improvement projects including:

- It displays large amounts of data that are difficult to interpret in tabular form.
- It illustrates quickly the underlying distribution data revealing the central tendency and variability of a data set.

Basic steps

1. Collect at least 50–125 data points for establishing a representative pattern.
2. Subtract the smallest individual value from the largest in the data set.
3. Divide this range by 5, 7, 9 or 11 depending on the number of data points. As a rough guide take the square root of the total number of data points and round it to the nearest integer. For example, for 50 data points divide the range by 7 and 125 data points by 11.
4. The resultant value determines the interval of the sample. It should be rounded up for convenience.
5. Calculate the number of data points in each group or class.
6. Plot the histogram with the intervals in the *x*-axis and the frequency of occurrence on the *y*-axis.
7. Clearly label the histogram.
8. Interpret the histogram related to centring, variation and shape (distribution).

Worked-out example

The following example is taken from Schmidt et al. (1999, pp. 135–137).

Table 5.1 shows the data points of miles per gallon data of a car pool.

Table 5.1 Data points of miles per gallon data of a car pool

18	16	30	29	28	21	17	41	8	17
32	26	16	24	27	17	17	33	19	18
31	27	23	38	33	14	13	26	11	28
21	19	25	22	17	12	21	21	25	26
23	20	22	19	21	14	45	15	24	34

For 50 points data set we would use $\sqrt{50} = 7.06 \rightarrow 7$ classes. Since our smallest data point is 8 and the largest is 45, the interval should be $(45 - 8)/7 = 5.3$. We round it up to make it 6. In Table 5.2 we place individual data in each class as shown.

Next we construct the histogram as shown in Figure 5.4. We see the most values are between 12 and 30 mpg with 18–24 mpg representing the median. The data are also skewed to the right by some high mileage values.

Table 5.2 Placing data in class ranges

Class number	Class range	Values	Frequency of occurrence
1	6–12	8, 11	2
2	12–18	16, 16, 17, 17, 14, 12, 14, 17, 17, 13, 15, 17	12
3	18–24	18, 21, 23, 19, 20, 23, 22, 22, 19, 21, 21, 21, 21, 19, 18	15
4	24–30	26, 27, 25, 29, 24, 28, 27, 26, 25, 24, 28, 26	12
5	30–36	32, 31, 30, 33, 33, 34	6
6	36–42	38, 41	2
7	42–48	45	1

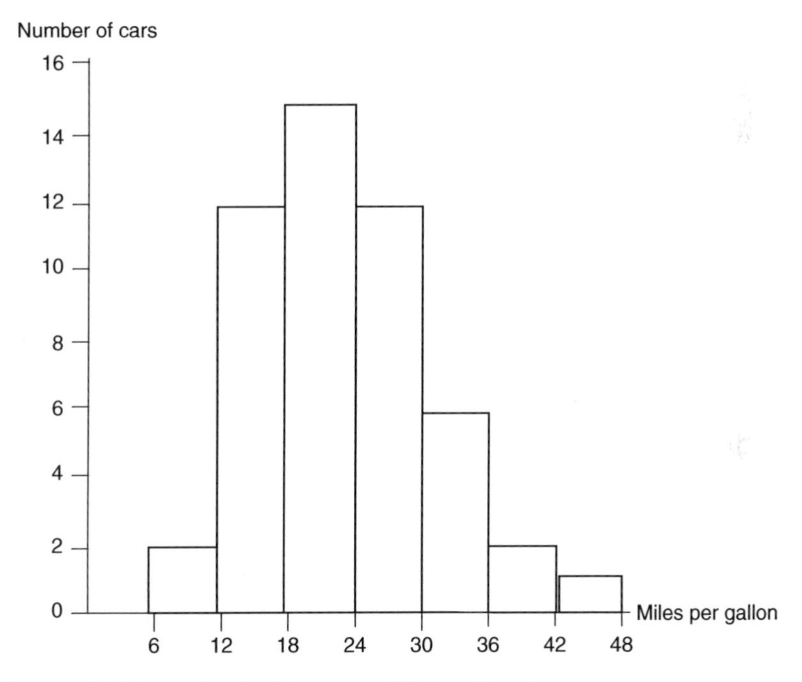

Figure 5.4 Example of a histogram (© Ron Basu).

Training requirement

Although a histogram can lead to complex statistical analysis, its basic principles are relatively simple. Most members of the project team are likely to have some familiarity with and knowledge of histograms. However, a 1-hour session of classroom training should be valuable to establish the methodology, especially related to the range, class and intervals.

Final thoughts

A histogram gives a pictorial presentation of how the data is distributed. It is easy to understand and use and it should be regarded as an essential tool of continuous improvement.

M3: Run Charts

Definition

A Run Chart is a graphical tool to allow a team to study observed data for trends over a specified period of time. It is basically a simple line graph of *x*- and *y*-axes.

Application

A Run Chart has a wide range of applications to detect trends, variation or cycles. It allows a team to compare performances of a process before and after the implementation of the solution. The application areas include sales analysis, forecasting, performance reporting and seasonality analysis.

Basic steps

1. Select the parameter and time period for measurement.
2. Collect data (generally 10–20 data points) to identify meaningful trends:
   a. *x*-axis for time or sequence cycle (horizontal)
   b. *y*-axis for the variable parameter that you are measuring (vertical).
3. Plot the data in a line graph along the *x*- and *y*-axis.
4. Interpret the chart. If there are no obvious trends then calculate the average value of the data points and draw a horizontal line at the average value.

Worked-out example

The following table shows the operational efficiency (%) of a packaging machine:

x	Jan	Feb	Mar	Apr	May	Jun	Jul	Aug	Sept	Oct	Nov	Dec
y	55	60	45	40	65	60	65	30	60	65	60	55

The Run Chart for the above data is shown in Figure 5.5.

Training requirement

The understanding and application of a Run Chart is so simple that it is not essential to conduct a classroom training specifically for it. The team members should be able to use a Run Chart after a preliminary briefing, in a

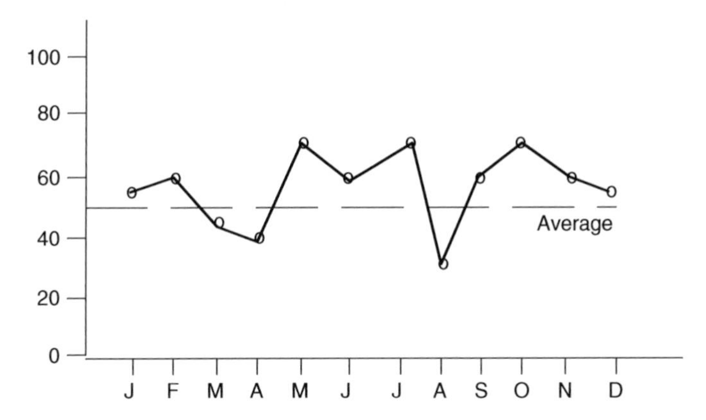

Figure 5.5 Example of a Run Chart (© Ron Basu).

practical application. The training process involving a Run Chart is usually included in a training programme for continuous improvement tools.

Final thoughts

Due to the fact that it could be seen as encompassing simplicity to a fault, a danger in using a Run Chart is the inclination to interpret every variation as significant. For any statistically significant result, Control Charts or Regression Analysis should be more appropriate. We recommend the use of a Run Chart for detecting a visual trend only and identifying areas of further analysis.

M4: Scatter Diagrams

Definition

A Scatter Diagram is a plot of points to study and identify the possible relationship between two variables, characteristics or factors. The knowledge provided by a Scatter Diagram can be enhanced more accurately by Regression Analysis.

Application

A Scatter Diagram is used, as an initial step before Regression Analysis, to show in simple terms if the variables are associated (a linear pattern) or unrelated (non-linear random pattern). Analysis should investigate the scatter of the plotted points and if some linear or non-linear relationship exists between two variables. In this the Scatter Diagram is very useful for diagnosis and problem solving.

A Scatter Diagram is often used in a follow-up to a Cause and Effect Diagram to detect if there are more than two variables between causes and the effect.

Basic steps

1. Collect paired samples of data (at least 10) of two variables that may be related.
2. Draw the x- and y-axes of the diagram.
3. Plot the data points on the diagram.
4. If some values are repeated, circle those points depending on the number of times they are repeated.
5. Interpret the results. The Scatter Diagram does not predict cause and effect relationships. It can, however, indicate a possible positive or negative correlation or no correlation between two variables.

Worked-out example

The following example is taken from Schmidt et al. (1999, pp. 154–156).

Table 5.3 shows a set of paired data points of two variables.

Table 5.3 A set of paired data points of two variables

Observations	Weight (lb)	Miles per gallon (mpg)
1	3000	18
2	2800	21
3	2100	32
4	2900	17
5	2400	31
6	3300	14
7	2700	21
8	3500	12
9	2500	23
10	3200	14

We plot the above data with weight as the x-axis (cause) and mpg as the y-axis (effect), as an example of a Scatter Diagram, as shown in Figure 5.6.

It appears that a negative correlation exists between two variables, because when one variable (weight) gets bigger then the other variable (mpg) becomes smaller.

Training requirements

It is important that training of Regression Analysis is conducted together with that of Scatter Diagrams. Although the topic of Scatter Diagrams may not require a classroom session on its own, their link with Regression Analysis should be understood clearly.

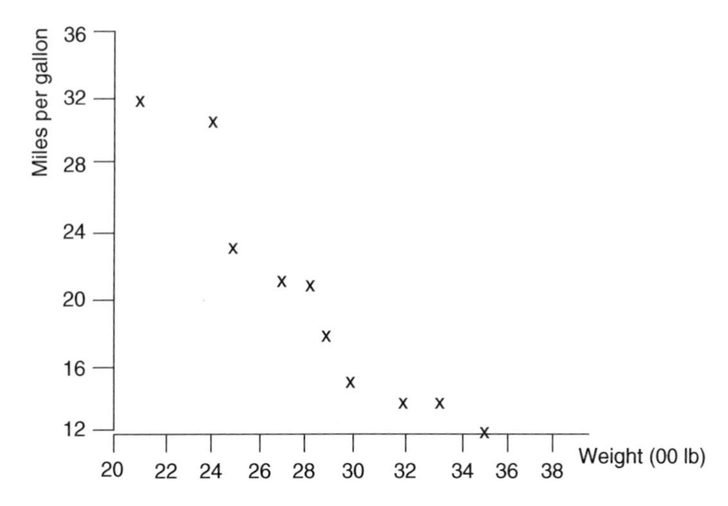

Figure 5.6 Example of a Scatter Diagram (© Ron Basu).

It is also useful to have a few minutes' training to assess whether the two variables have possible positive correlation, no correlation or a possible negative correlation.

Final thoughts

Scatter Diagrams are easy to understand and apply. We recommend Scatter Diagrams as the first step to study the relationship of two variables before embarking on more complex Regression Analysis.

M5: Cause and Effect Diagrams

Definition

The Cause and Effect Diagram is a graphical representation of potential causes for a given effect.

Since it was first used by Ishikawa, this type of illustration is also known as an Ishikawa diagram. In addition, it is often referred to as a 'fishbone' diagram due to its skeletal appearance.

The purpose of the diagram is to assist in brainstorming and enabling a team to identify and graphically display, in increasing detail, the root causes of a problem.

Application

The Cause and Effect Diagram is arguably the most commonly used of all quality improvement tools. The 'effect' is a specific problem and is considered to constitute the head of the diagram. The potential causes and sub-causes of the problem form the bone structure of the skeletal fish.

They are typically used both during the measurement and analysis phase of the project. Their wide application area covers Six Sigma teams, TQM teams or Continuous Improvement teams as part of brainstorming exercises to identify the root causes of a problem and offer solutions. It focuses the team on causes, rather than symptoms.

There are a number of variants in the application of a Cause and Effect Diagram. The two most common types are the 6M Diagram and Cause and Effect Diagram Assisted by Cards (CEDAC). Let us examine each of these terms a little more closely.

In a 6M Diagram, the main bone structure or branches typically consist of the self-explanatory '6Ms':

- Machine
- Manpower
- Material
- Method
- Measurement
- Mother Nature (Environment).

A CEDAC Diagram works slightly differently. A blank, highly visible fishbone chart is displayed in a meeting room. Every member of the team must post both potential causes and solutions on a card (or Post-It™ notes) considering each of the categories.

A CEDAC also consists of two major formats:

1. Dispersion Analysis Type
2. Process Classification Type.

Firstly, the Dispersion Analysis Type is used usually after 6M or CEDAC Diagrams have been completed. The major causes identified are then treated as separate branches and their sub-causes are identified.

However the Process Classification Type uses the major steps of the process in place of the major cause categories. This form is usually used when the problem encountered cannot be isolated to a single department. Each stage of the process is then analysed by using a 6M or CEDAC approach.

A typically sequence of Cause and Effect Diagrams for a complex problem could be:

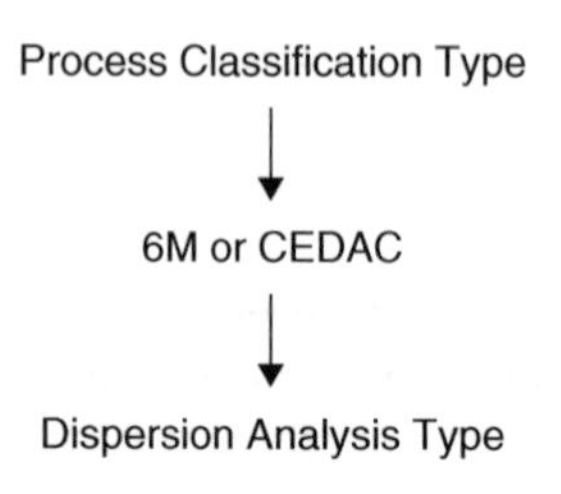

Basic steps

1. Select the most appropriate cause and effect format. If the problem can be isolated to a single section or department choose either a 6M (small team) or a CEDAC (large team) approach.
2. Define with clarity and write the key effect of the problem in a box to the right-hand side of the diagram.
3. Draw a horizontal line from the left-hand side of the box. Draw main branches (fishbones) of the diagram after agreeing the main categories (e.g. 6M) of causes.
4. Brainstorm for each category the potential sub-causes affecting the category.
5. List the sub-causes of each category in a flipchart. In a CEDAC approach these would be a collection of Post-It™ notes for each category.
6. Rank the sub-causes in order of importance by a group consensus (or multi-voting) and select up to six top sub-causes for each category.
7. Construct the diagram by posting the top sub-causes in each category. These are the 'root causes'.
8. Decide upon further Dispersion Analysis or gather additional data needed to confirm the root causes.
9. Develop solutions and improvement plans.

Worked-out examples

The following example is taken from Basu and Wright (2003, pp. 29–30).

Consider the situation where customers of a large international travel agency sometimes find that when they arrive at their destination, the hotel has no knowledge of their booking.

In this case, to get started you might begin with 'Hotel Not Booked' as the effect and the 6Ms (Machine, Manpower, Method, Material, Measurement and Mother Nature) as the causes. When the sub-causes are further investigated the diagram may look like that shown in Figure 5.7.

The diagram points out that many sub-causes including training, e-mail systems and customer feedback appear to be worthwhile areas to follow up.

Training requirements

As Cause and Effect Diagrams are essential to a quality improvement programme, it is important that members of the team receive at least 1 hour of hands-on training. An effective method of training is by participating in a brainstorming exercise where a Cause and Effect Diagram is developed. It is more important to identify the root causes than to assign the cause in a specific category. For instance, in the example given in Figure 5.7, 'clear instructions' could also be grouped under Manpower or 'checklist' could be assigned to the category of Measurement.

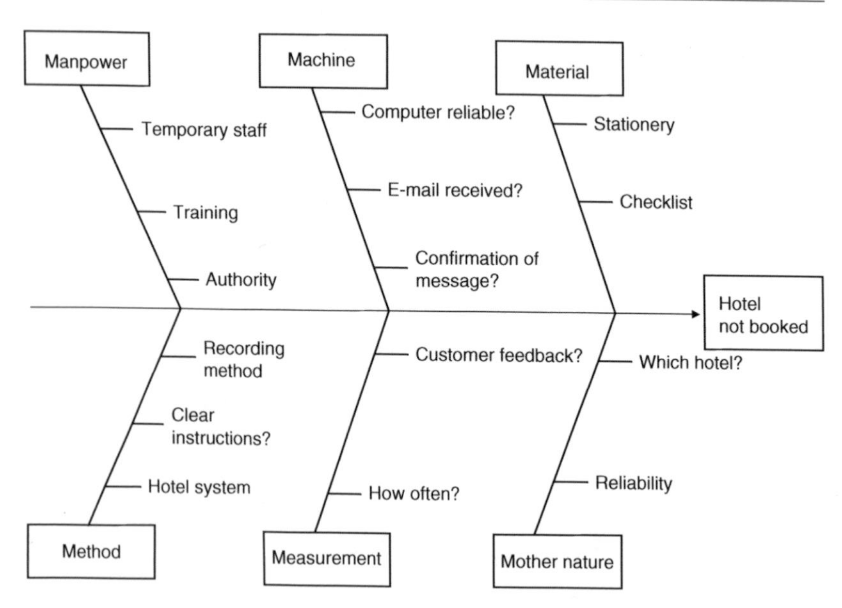

Figure 5.7 The Cause and Effect Diagram (© Ron Basu).

Final thoughts

Cause and Effect Diagrams are invaluable in brainstorming the root causes for a given problem. We strongly recommend their application on as wide a basis as possible.

M6: Pareto Charts

Definition

A Pareto Chart is a special form of bar chart that rank orders the bars from highest to lowest in order to prioritise problems of any nature.

It is known as 'Pareto' after a nineteenth century Italian economist Wilfredo Pareto who observed that 80% of the effects are caused by 20% of the causes: 'the 80/20 rule'.

Applications

Pareto Charts are applied to analyse the priorities of problems of all types, e.g. sales, production, stock, defects, sickness, accident occurrences, etc. The improvement efforts are directed to priority areas that will have the greatest impact.

There are usually two variants in the application of Pareto Charts. The first type is the standard chart where bar charts are presented in descending order. The second type is also known as the 'ABC Analysis' where:

- Cumulative % values of causes are plotted along the *x*-axis.
- Cumulative % values of effects are plotted along the *y*-axis.
- 80% of the effects with corresponding % of causes are grouped as 'A' category.
- 80–96% of the effects and corresponding causes are 'B' items.
- The remaining values are 'C' category.

Basic steps

The following steps apply for the preparation of a Pareto Chart.

1. Identify the general problem (e.g. IC Board Defects) and its causes (e.g. Soldering, Etching, Moulding, Cracking, Other).
2. Select a standard unit of measurement (e.g. Frequency of Defects or Money Loss) for a chosen time period.
3. Collect data for each of the causes in terms of the chosen unit of measurement.
4. Plot the Pareto Chart with causes along the *x*-axis and the unit of measurement along the *y*-axis. The causes are charted in descending order of values from left to right.
5. Analyse the graph and decide on the priority for improvement.

The following steps apply for the preparation of an ABC Analysis:

1. Decide on the causes (e.g. number of customers) and effect (e.g. sales) of the problem areas that you want to prioritise.
2. Select a standard unit of measuring the effect (e.g. $ for sales values).
3. Collect data for each cause (e.g. customer) and the corresponding effect (e.g. sales in $).
4. Rank the cause according to the value of the effects (e.g. customers in descending order of $ sales).
5. Calculate the cumulative % values of the causes and effects.
6. Plot the cumulative % values of the causes (e.g. number of customers) along the *x*-axis and the cumulative % values of effects (e.g. sales in $) along the *y*-axis.
7. Identify A, B and C categories (e.g. A for 80% of sales, B for 80–96% of sales and C for the remainder).
8. Analyse the graph and decide on the priority for improvement.

Worked-out examples

The following example of a Pareto Chart is taken from Schmidt et al. (1999, p. 144).

The major defects identified during the manufacture of Integrated Circuit Boards for a given period were given in Table 5.4.

Table 5.4 Major defects identified during the manufacture of Integrated Circuit Boards

Causes	Frequency	% Frequency
Soldering	60	40
Etching	40	27
Moulding	30	20
Cracking	15	10
Other	5	3
Total	150	100

We plot the causes along the *x*-axis and frequencies along the *y*-axis as shown in Figure 5.8.

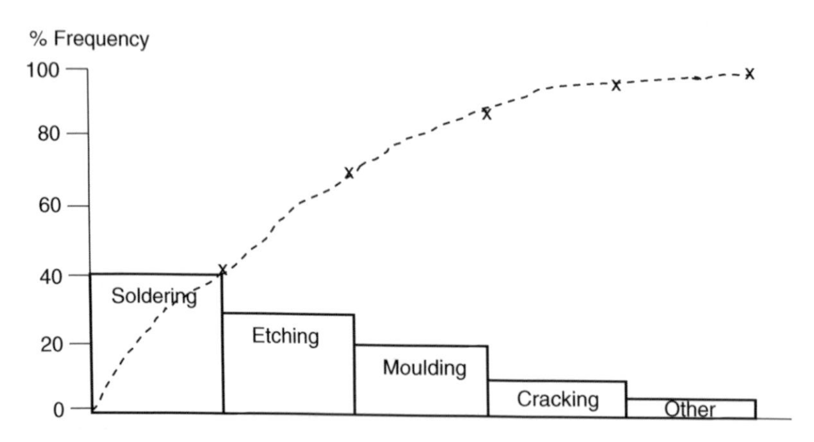

Figure 5.8 Pareto Chart (© Ron Basu).

An example of ABC Analysis for a set of inventory items is shown in Figure 5.9.

Training requirement

The basic principles of a Pareto Chart and ABC Analysis are easy to follow. The training for this tool of prioritising is usually included in the classroom training sessions related to the tools for measurement. A member of a team can be adept in the application of Pareto Charts after, say, 1 hour's practice on a few practical problems.

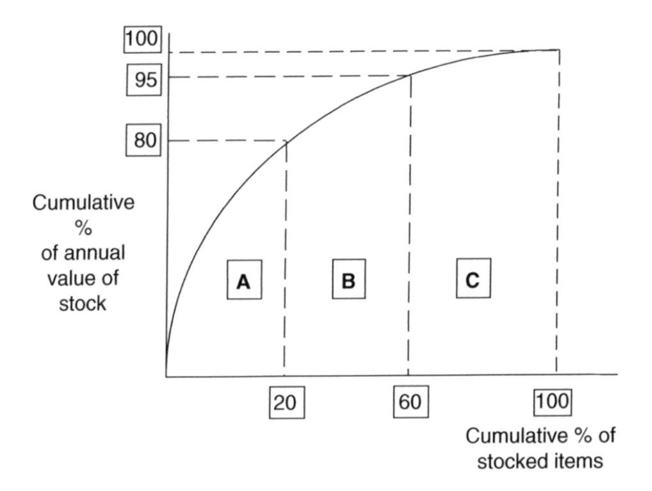

Figure 5.9 ABC Analysis.

Final thoughts

Pareto Charts are extremely useful in optimising efforts and 'going for gold'. They have the twin advantages of being both easy to understand and to apply. Thus we highly recommend the use of Pareto Charts for all quality improvement projects.

M7: Control Charts

Definition

A Control Chart consists of a graph with time on the horizontal axis and an individual measurement (such as mean or range) on the vertical axis.

A Control Chart is a basic graphical tool of Statistical Process Control (see Chapter 9) for determining whether a process is stable and also for distinguishing usual (or common) variability from unusual (special assignable) causes. Three control limits are drawn: the central line (CL), the lower control limit (LCL) and the upper control limit (UCL). The points above the UCL or below the LCL indicate a special cause. If no signals occur the process is assumed to be under control, i.e. only common causes of variation are present.

Application

Control Charts can be used for examining a historical set of data and also for current data. The current control based on current data underpins a feedback control loop in the process.

There are many good reasons why Control Charts have been applied successfully in both quality control and improvement initiatives since Walter Shewhart first introduced the concept at the Bell Laboratories in the early 1930s.

First, Control Charts establish what is to be controlled and force their resolution. Secondly, they focus attention on the process rather than on the product. For example, a poor product can result from an operator error, but a poor production process is not capable of meeting standards on a consistent basis. The third factor is that they comprise a set of prescribed techniques that can be applied by people with appropriate training in a specified manner.

For many probability distributions, most of the probability is within three standard deviations of the mean. So μ and σ are respectively the stable process mean and standard deviation, then

$$\begin{aligned} \text{CL} &= \mu \\ \text{UCL} &= \mu + 3\sigma \\ \text{LCL} &= \mu - 3\sigma \end{aligned}$$

When the mean (μ_s) and standard deviation (σ_s) values are calculated from a sample of n then

$$\begin{aligned} \mu_s &= \mu \\ \sigma_\sigma &= \sigma/\sqrt{n} \end{aligned}$$

There are two types of Control Charts. A variable chart is used to measure individual measurable characteristics, whereas an attributes chart is used for go/no-go types of inspection.

The x-bar chart (also called the mean chart), the s-chart (also called the standard deviation chart) and the R-chart (also called the range chart) are used to monitor continuous measurement or variable data.

The stable Control Charts are used to determine Process Capability, i.e., whether a process is capable of meeting established customer requirements or specifications.

Basic steps

1. Choose the quality characteristic to be charted. A Pareto analysis is useful to identify a characteristic that is currently experiencing a high number of non-conformities.
2. Establish the type of Control Chart to ascertain whether it is a variable chart or an attribute chart.
3. Choose the subgroup or sample size. For variable charts, samples of 4–5 are sufficient, whereas for attribute charts samples of 50–100 are often used.
4. Decide on a system of collecting data. The automatic recording of data by a calibrated instrument is preferable to manually recorded data.
5. Calculate the mean and standard deviation and then calculate the control limits.
6. Plot the data and control limits on a Control Chart and interpret results.

Worked-out example

The following example illustrates the construction of variable Control Charts based on the data of packing cartons of a morning shift. The data in Table 5.5 is taken from Ledolter and Burnill (1999).

Table 5.5 Packing cartons in a morning shift

Reading number	Measurements				Average	Standard deviation	Range
1	25.1	25.5	25.0	25.1	25.175	0.222	0.50
2	24.8	25.2	25.1	24.9	25.000	0.183	0.40
3	25.1	25.2	25.2	25.2	25.175	0.050	0.10
4	25.1	25.4	24.8	25.0	25.075	0.250	0.60
5	25.2	24.7	24.9	25.3	25.025	0.275	0.60
6	25.2	25.2	25.0	25.1	25.125	0.096	0.20
7	25.2	25.2	25.2	25.3	25.225	0.050	0.10
8	25.2	25.1	25.3	25.0	25.150	0.129	0.30
9	24.9	25.1	25.2	24.8	25.000	0.183	0.40
10	25.1	25.1	25.3	25.4	25.225	0.150	0.30
Average					25.118	0.159	0.35

Hence

$$
\begin{aligned}
\text{CL} &= \mu_s = 25.118 \\
\sigma_s &= \sigma/\sqrt{n} = 0.159/\sqrt{10} = 0.159/3.163 = 0.05 \\
\text{UCL} &= \mu_s + 3\mu_s = 25.118 + 3 \times 0.05 = 25.27 \\
\text{LCL} &= \mu_s - 3\sigma_s = 25.118 - 3 \times 0.05 = 24.97
\end{aligned}
$$

The control limits and the data are plotted as shown in Figure 5.10.

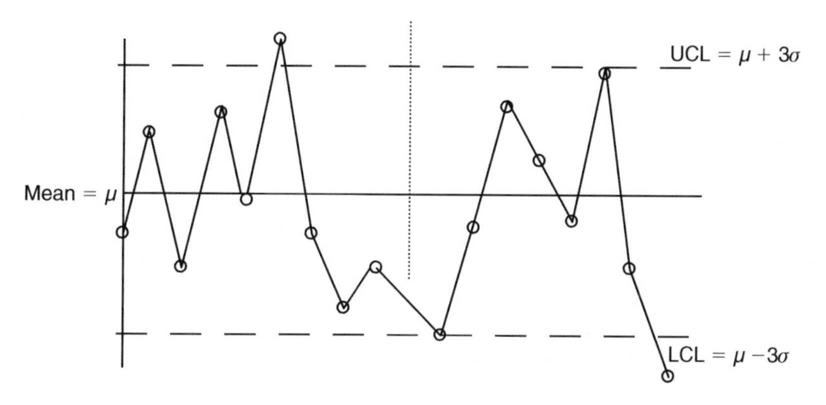

Figure 5.10 Control Chart.

Similar charts are drawn for the standard deviation and the range of the data. The control limits are constructed in such a manner that we expect approximately 99.7% of all data to fall between them. In Figure 5.10 all data are within the control limits, indicating that the process is stable.

Training requirement

The construction and interpretation of Control Charts require a good understanding of statistical process control. The concepts of variable data versus attribute data, common causes versus special causes and control limits are essential to the effective application of Control Charts. Hence a few hours of classroom training is recommended before the members start the application of this tool. Care should be taken not to confuse a Control Chart with a Run Chart.

Final thoughts

Control Charts are useful to identify data and their causes outside the control limits. However, nothing will change just because you charted it. You need to do something and eliminate the causes.

M8: Flow Process Charts

Definition

A Flow Process Chart is a symbolic representation of a physical process linked to correspond to the sequence of operation.

It was originally introduced as a classical industrial engineering tool with five symbols (collectively known as activities) as follows:

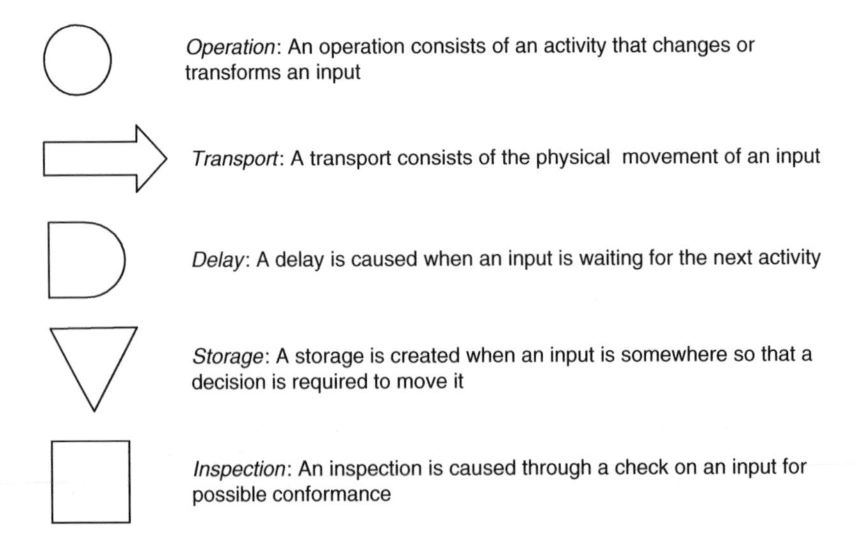

Application

Flow Process Charts have been used extensively in manufacturing and supply chain operations to identify:

- The hierarchical structure of operations
- The sequence of activities
- Non-value added activities.

Over the years, many specialised forms of flow charts have evolved to analyse the hierarchical structure and sequence of activities. Two such specialised derivatives are Flow Diagrams (for data processing) and Process Mapping (for process sequencing). However, the classical Flow Process Charts are still being applied to identify non-value added activities: Their applications have also been extended to service and transaction activities. A more recent application of Flow Process Charts has been in the analysis of Lean Processes (Basu and Wright, 2003, p. 71) where seven forms of wastes or 'mudas' (non-value added activities) have been defined:

1. Excess production
2. Waiting
3. Transportation
4. Motion
5. Process
6. Inventory
7. Defects.

Traditionally, there are two formats for Flow Process Charts: a pre-printed format and a descriptive arrangement. In the pre-printed format, each activity is recorded in the form, the corresponding symbols are marked and then joined in sequence by a line. This is the classical industrial engineering format where the number of activities for the present and proposed methods is compared.

In the descriptive setup, each activity is charted in sequence and represented by process symbols. This chart provides a systematic start to analyse the process and does not require any extra stationery.

Basic steps

1. Select the process for analysis. Start with a high level block diagram to gain a broad process view.
2. Identify the scope of the Flow Process Chart by defining physical and functional boundaries.
3. Develop the sequence of process activities. Capture activity details as they really are, rather than what could be, according to written procedures.
4. Develop the Flow Process Chart by assigning the appropriate symbols for each activity.

5. Identify non-value added activities and opportunities for improvement.
6. Develop the Flow Process Chart for the improved process and quantify that improvement.

Worked-out example

The following example is taken from Kolarik (1995, p. 212).

Consider the development of a Flow Process Chart for a cattle feed manufacturing process. Figure 5.11 shows a high level 'block diagram' of the process.

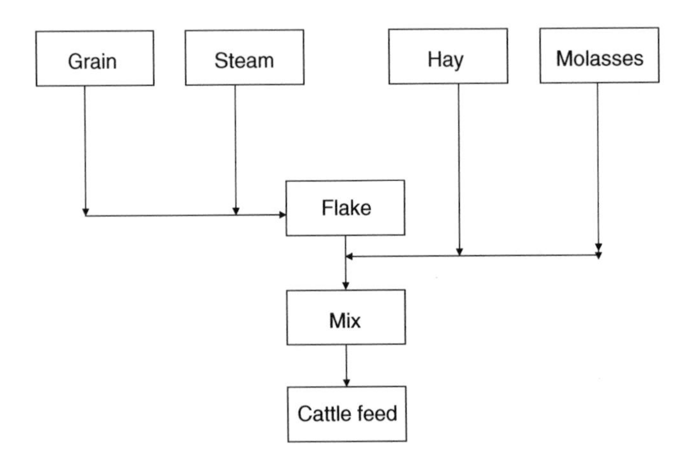

Figure 5.11 Block Diagram (© Ron Basu).

We scope the Flow Process Chart for 'grain' related activities only for the purposes of this example. The details of this process are shown in the Flow Process Chart in Figure 5.12.

The chart in Figure 5.12 identifies the non-value added activities of transport, inspection and storage, and points towards a direct loading of grains to the steam chamber. This would lead to the feasibility of the quality assurance of grains before loading and an automated weight control system so that grains can be blown directly to the steam chamber.

Training requirements

The basic principles of the Flow Process Chart are fundamental and can be applied consistently. This makes the training process relatively simple and thus team members can use Flow Process Charts effectively after 1 hour's practice on a practical problem. This is a manual process and does not require any computer software like Process Mapping.

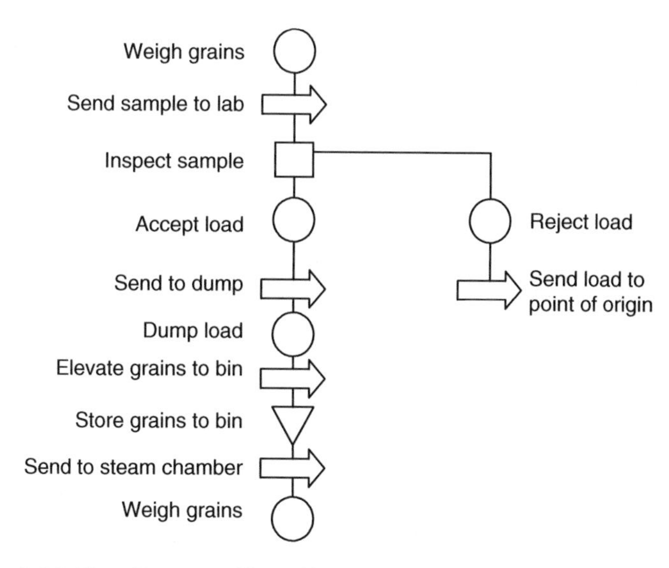

Figure 5.12 Flow Process Chart (© Ron Basu).

Final thoughts

With the advent of computerised Process Mapping facilities, the use of Flow Process Charts is not fashionable any more. However, it is still an effective manual tool to identify non-value added activities. We recommend its application in a simple process or as part of a more complex course of action.

M9: Process Capability Measurement

Definition

Process Capability is the statistically measured inherent reproducibility of the output (product) turned out by a process.

A commonly used measure of process capability is given by the capability index (C_p):

$$C_p = \frac{\text{USL} - \text{LSL}}{6\sigma}$$

where USL and LSL are upper and lower specification limits and σ is the standard deviation.

The C_p index measures the potential or inherent capability of process. The C_{pk} index measures the realised process capability relative to the actual operation and is defined as

$$C_{pk} = \text{minimum}\left(\frac{\mu - \text{LSL}}{3\sigma}, \frac{\text{USL} - \mu}{3\sigma}\right)$$

If $C_{pk} > 1$, we declare that the process is capable, and if $C_{pk} < 1$, then we declare that the process is incapable.

C_{pk} is a more practical measure of capability than C_p.

Application

A customer requires that specifications are given in terms of a target value, a LSL and USL. They are the 'tolerances' of the specification. The process capability index determines the reliability of these tolerances to be delivered by the process used by the supplier.

The process capability information serves many purposes including:

- Selection of competing processes or equipment that best satisfies the specification.
- Predicting the extent of variability that processes could exhibit to establish realistic specification limits.
- Testing the theories of cause and effect during quality improvement programmes.

Motorola introduced the concept of Six Sigma as a statistical way of measuring total customer satisfaction. Given that the process is a Six Sigma process, we know that USL $-$ LSL $= 6\sigma + 6\sigma = 12\sigma$. Hence

$$C_p = \left(\frac{\text{USL} - \text{LSL}}{6\sigma} = \frac{12\sigma}{6\sigma} \right) = 2.0$$

Another metric, DPMO (defects per million opportunities), is also used to assess the performance of a process. DPMO is represented by the proportion of area outside the specification limit multiplied by one million. For example, if 3.5% of the area is outside specification limits, then

$$\text{DPMO} = 1{,}000{,}000 \times \frac{3.5}{100} = 35{,}000$$

The three metrics DPMO, C_p and C_{pk} all given numerical values that indicate how well a process is doing with respect to these specification limits.

Basic steps

1. Using a stable Control Chart, determine the process grand average (χ), the average (R) and process standard deviation (s).
2. Determine the USL and the LSL based upon customer requirements.
3. For a stable process, assume that the values of the sample mean and standard deviation are the same as the corresponding values of the population, i.e. $\sigma = s$ and $\mu = \chi$.
4. Calculate the process capability indices C_p and C_{pk} by using the formulae:

$$C_p = \frac{\text{USL} - \text{LSL}}{6\sigma}$$

$$C_{pk} = \text{Minimum}\left(\frac{\mu - \text{LSL}}{3\sigma}, \frac{\text{USL} - \mu}{3\sigma}\right)$$

Figure 5.13 illustrates three states of the potential process capability for $C_p < 1$, $C_p = 1$ and $C_p > 1$.

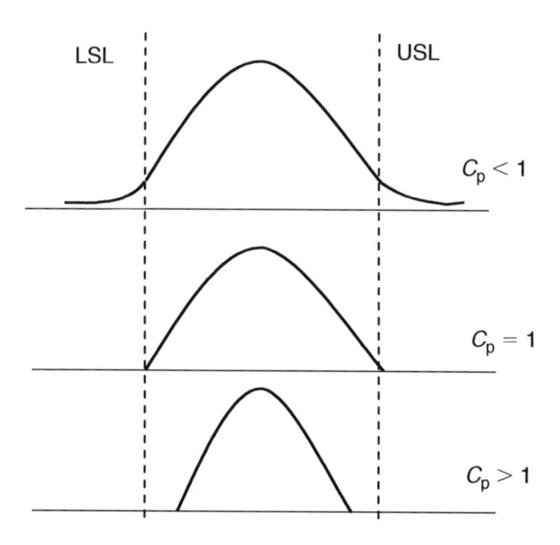

Figure 5.13 Process Capability.

Worked-out example

Consider the graph in Figure 5.14.

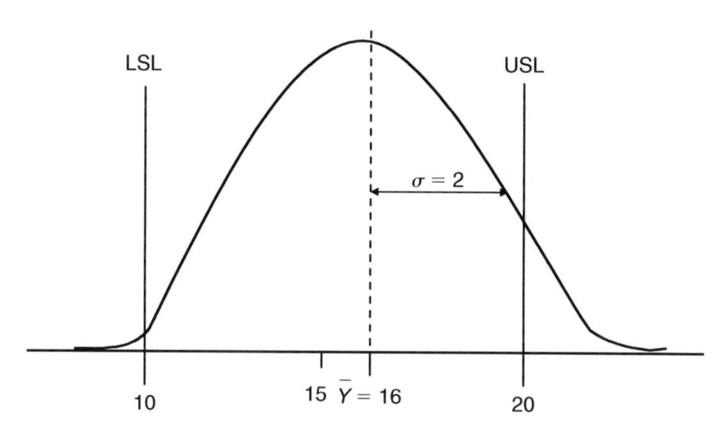

Figure 5.14 A distribution graph.

From the graph we get

$$\mu = \chi = 16$$
$$\sigma = s = 2$$
$$\text{USL} = 20$$
$$\text{LSL} = 10$$

Therefore we calculate

$$C_{\text{p}} = \frac{\text{USL} - \text{LSL}}{6\sigma} = \frac{20 - 10}{6 \times 2} = \frac{10}{12} = 0.83$$

$$C_{\text{pk}} = \text{Minimum}\left(\frac{\mu - \text{LSL}}{3\sigma}, \frac{\text{USL} - \mu}{3\sigma}\right)$$
$$= \text{Minimum}\left(\frac{16 - 10}{6}, \frac{20 - 16}{6}\right)$$
$$= \text{Minimum}\left(\frac{2}{3}, 1\right) = 0.67$$

Training requirements

The application of Process Capability requires a good understanding of Statistical Process Control and the relationship between the sample data and the population data. It is important to recognise that USL and LSL are based on customer requirements and are different from the corresponding values of the Control Chart.

A few hours of classroom training for Process Capability along with Control Charts and Statistical Process Control is recommended for specialist members of the team. In a Six Sigma programme, Black Belts go through a formal training schedule in the Process Capability.

Final thoughts

Process Capability is a useful tool in determining whether a process can deliver the customer specifications. However, we recommend that its use and interpretation should be handled by Black Belts or an adequately trained member.

6

Tools for analysis

All intelligent thoughts have already been thought; what is necessary is only to try to think them again.

— Johann Wolfgang Goethe

Introduction

Once the project is understood and defined in the 'Define' stage, and then the baseline performance has been documented and validated at the 'Measure' stage to ascertain that there is a real opportunity, it is now time to perform an in-depth analysis of the process. At this 'Analysis' stage, tools and techniques are applied to identify and validate the root causes of problems. The objective is to identify all possible sources of variation in the process and distinguish between special and common causes of variation. Having got to the root causes of the problem, the business cause of the Measure stage can be updated with more accurate data. The data collected in the Measure stage are examined to generate a prioritised list of sources of variation.

The key deliverables of the Analysis phase are:

1. *A prioritised list of variables*: a prioritised list of important sources of variation (particularly special causes) that affect the process output.
2. *Quantified financial opportunity*: the financial benefit expected from the completion of the project.

The important tools for analysis should include:

A1: Process Mapping
A2: Regression Analysis
A3: RU/CS Analysis
A4: SWOT Analysis
A5: PESTLE Analysis
A6: The Five Whys

A7: Interrelationship Diagram
A8: Overall Equipment Effectiveness
A9: TRIZ: Innovative Problem Solving

During the Analyse stage, some of the tools from the Define and Measure stages are revisited, in particular:

M4: Scatter Diagram
M5: Cause and Effect Diagram
M6: Pareto Chart
M7: Control Charts

The Analyse stage also depends heavily on advanced techniques including SPC, FMEA and DOE.

A1: Process Mapping

Definition

Process Mapping is a tool to represent a process by a diagram containing a series of linked tasks or activities which produce an output.

It is a further development of a Flow Diagram by using computer software so that the user can link quickly the activities and drill down to gain a more detailed picture.

Application

With the advent of well supported software (e.g. 'Control' by Enigma Ltd.), Process Mapping is becoming a way of life for analysing a process or an organisation.

A process map does not use symbols like a Flow Process Chart or a Flow Diagram. Only boxes and arrows are used and different colours are often applied to identify types of activities (e.g. non-value added or value added). There are several benefits of applying Process Mapping including the following. Process Mapping means that the team:

- Can clarify what is happening within an organisation
- Can simulate what should be happening
- Can show a process at various levels of detail
- Can allocate ownership of each activity and promote teamwork
- Can reflect the end-to-end process and its visibility
- Can add resources, costs, volumes and duration to build up sophisticated cost models
- Can identify how the performance of this process can be measured.

Basic steps

1. Decide on the organisation, function or process for analysis.
2. Agree higher level functions and their relationships.
3. Agree on input, process and output for each activity.
4. Construct a process diagram for selected functions.
5. Validate the process diagram with stakeholders.
6. Add resources, costs, volumes and duration if required.
7. Apply a 'what-if' analysis and simulation to achieve sustainable process improvement.

Worked-out example

The following example is taken from the demonstration package of the 'Control' software (courtesy: Enigma Ltd., Oxford).

Consider a case example of dealing with faults in a computer network. A high level process map is shown in Figure 6.1.

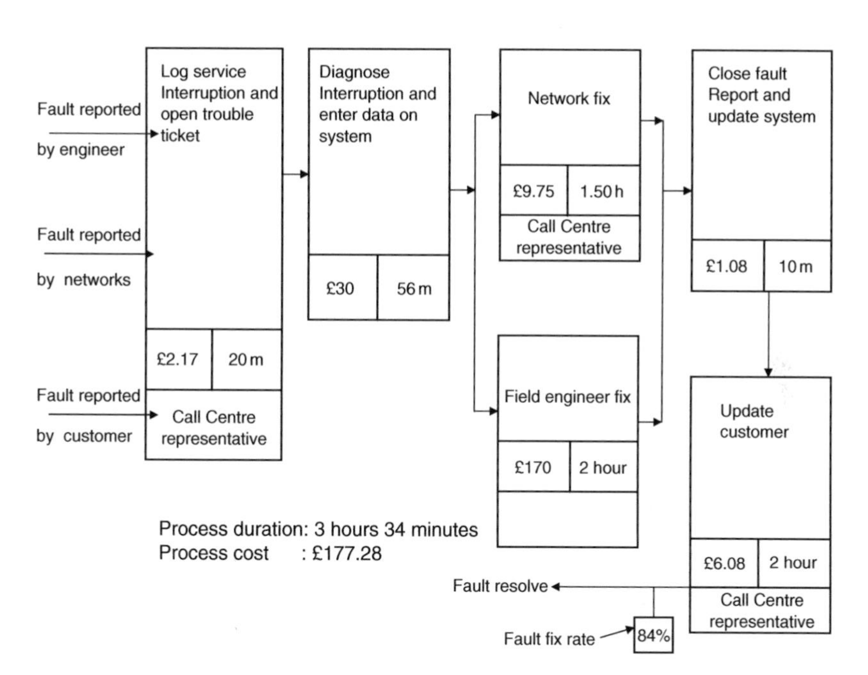

Figure 6.1 Process Mapping (© Ron Basu).

Training requirements

Although the basic principles of Process Mapping are simple and logical, it is important that users receive at least 1 day of hands-on training in the chosen software for Process Mapping. It is also useful to gain a good understanding of the 'what-if' simulation processes.

Final thoughts

Process Mapping has become a very useful computer aided tool for process improvement. However, it should be used for process mapping's sake. Process maps do not – in isolation – change individual behaviour.

A2: Regression Analysis

Definition

Regression Analysis is a tool to establish the 'best fit' linear relationship between two variables.

The knowledge provided by the Scatter Diagram is enhanced with the use of regression.

Application

The topic of Regression Analysis is usually studied at school in algebra lessons where different methods of 'curve fitting' are considered. Two common methods are:

1. Method of intercept and slope
2. Method of least square.

In a practical business environment, the team members normally resort to drawing an approximate straight line by employing their visual judgement. Sometimes they use the 'method of intercept and slope.' Both of these practices are the estimated 'best fit' relationship between two variables. The reliability of such estimates depends on the degree of correlation that exists between the variables.

Regression Analysis is used not only to establish the equation of a line but also to provide the basis for the prediction of a variable for a given value of a process parameter. The Scatter Diagram on the other hand does not predict cause and effect relationships. Given a significant co-relation between the two variables, Regression Analysis is very useful tool enabling one to extend and predict the relationship between these variables.

Basic steps

We have considered the 'method of intercept and slope' for developing the basic steps as follows (courtesy: Moroney, 1973, p. 284):

1. Consider the equation of $y = mx + c$, where m is the slope, c is the intercept and x and y are the two variables.
2. In the equation of $y = mx + c$, substitute each of the pairs of values for x and y and then add the resulting equations.

3. Form a second similar set of equations, by multiplying through each of the equations of Step 2 by its co-efficient of m. Add this set of equations.
4. Steps 2 and 3 will each have produced an equation in m and c. Solve these simultaneous equations for m and c.
5. Plot the straight line graph for $y = mx + c$ for the calculated values of m and c.

Worked-out example

The following example is taken from Moroney (1973), pp. 278–285.

Consider an investigation is made into the relationship between two quantities y and x and the following values were observed:

y	5	8	9	10
x	1	2	3	4

The values are plotted as shown in Figure 6.2. Now we follow the basic steps to calculate m and c in the equation $y = mx + c$.

Substituting the observed values of x and y, the resulting equations are:

$$\begin{aligned} 5 &= m + c \\ 8 &= 2m + c \\ 9 &= 3m + c \\ 10 &= 4m + c \\ 32 &= 10m + 4c \end{aligned} \tag{6.1}$$

Multiplying each of the equations by its co-efficient of m, the resulting equations are:

$$\begin{aligned} 5 &= m + c \\ 16 &= 4m + 2c \\ 27 &= 9m + 3c \\ 40 &= 16m + 4c \\ 88 &= 30m + 10c \end{aligned} \tag{6.2}$$

We then solve simultaneously equations (6.1) and (6.2) for m and c.

We find that

$$\begin{aligned} &m = 1.6 \\ &c = 4 \\ &\text{Hence } y = 1.6x + 4 \end{aligned}$$

We calculate two pairs of values for x and y to draw a straight line as shown in Figure 6.2.

y	4	8.8
x	0	3

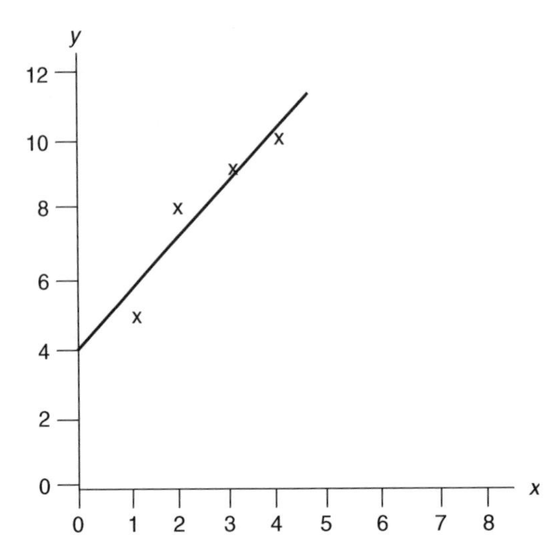

Figure 6.2 Regression Analysis (© Ron Basu).

Training requirements

The training of Regression Analysis is like brushing up on school algebra. A few classroom exercises for approximately 1 hour should be adequate for team members to prepare themselves for the practical application of Regression Analysis.

Final thoughts

Regression Analysis is a powerful tool to extend data from a Scatter Diagram. We recommend the application of Regression Analysis preferably by a team member who has competence and an interest in mathematics.

A3: RU/CS Analysis

Definition

The RU/CS (Resource Utilisation and Customer Service) Analysis is a simple tool to establish the relative importance of the key parameters of both RU and CS and to identify their conflicts.

Wild (2002) suggests the starting point of the RU/CS Analysis with the Operations Objective Chart as shown in Figure 6.3.

The relative importance of the key parameters for RU (i.e. Machines, Materials and Labour) and CS (i.e. Specification, Cost and Time) can be given a rating of 1, 2 or 3 (3 being the most important).

	Resource Utilisation			Customer Service		
	Machines	Materials	Labour	Specification	Cost	Time
Operation						

Figure 6.3 Operations Objective Chart.

Application

In any business or operation, a manager has to find a balance between two conflicting objectives. CS is of course the primary objective of the operation. For simplicity, three key parameters of CS are considered. These are Specification, Cost (or Price) and Timing. The customer expects the goods or service to be delivered according to acceptable standards, to be of an affordable price and that they arrive on time. However, the relative importance of Specification, Cost and Time could change depending on the market condition, competition and the desirability of demand. The second objective of the Operation Manager is to utilise resources to meet CS requirements. Given infinite resources, any system can provide adequate CS, but many companies have gone out of business in spite of possessing satisfied customers. Therefore, it is essential to provide an efficient use of resources. The RU/CS Analysis aims to point out the way forward to a balanced approach of 'effective' CS and 'efficient' RU.

An organisation in a normal condition will not aim to maximise all three parameters. Likewise, few organisations will aim to maximise the utilisation of all resources. Hence there is some room for adjustment and the Operations Manager must attempt to balance the parameters of these two basic objectives – RU and CS. The RU/CS Analysis is applied to examine the relative importance of the parameters to lead to a balanced solution of objectives.

Basic steps

1. Identify the key parameters of RU. An operation may have several types of resources as input (e.g. machine, facilities, labour, information, etc.). Choose three important resources.
2. For CS parameters, select Specification, Cost and Time. As discussed in Chapter 1, there are other dimensions of quality as perceived by customers; for the sake of simplicity only Specification has been chosen as the key parameter for the quality of service.
3. Draw two matrices for RU and CS showing the six parameters.
4. Allocate a rating of 1, 2 or 3 (3 being most important) to the parameters of both RU and CS. The ratings are influenced by internal processes for RU and external customer requirements for CS.
5. Rate separately what is actually achieved for each aspect of RU and CS.
6. Compare the two sets of figures (from Steps 4 and 5) and identify the shortfalls or misalignments.

7. Review the criticality of shortfall in CS and examine which resources are inhibiting CS performance.
8. Draw a combined RU/CS matrix, with the allocated ratings outlined in Step 4, and identify their conflicts. It is important to note that the high importance of Specification and Time will require a lower RU. The high importance of cost, on the other hand, will demand a lower price and this will require a higher RU. The tables in Figure 6.4 can be used as a ready reckoner to identify conflicts.
9. Having identified the conflicts, the next step is to examine the relative importance of each parameter in order to minimise conflicts.

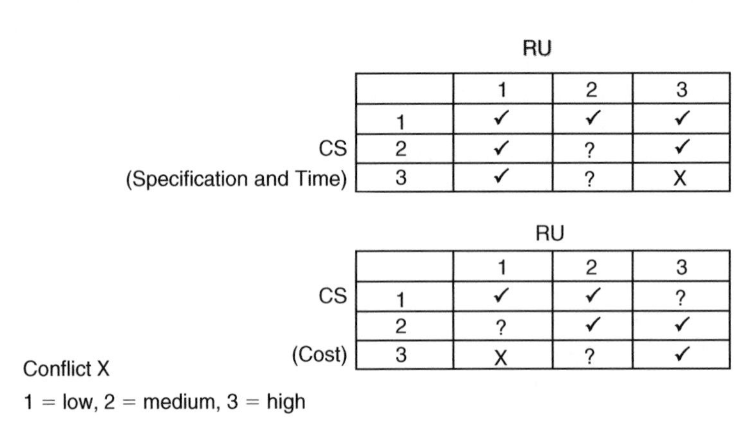

CS (Specification and Time)	RU 1	RU 2	RU 3
1	✓	✓	✓
2	✓	?	✓
3	✓	?	X

CS (Cost)	RU 1	RU 2	RU 3
1	✓	✓	?
2	?	✓	✓
3	X	?	✓

Figure 6.4 RU/CS conflicts.

Worked-out example

Consider a mail order company where customers are expecting good value for money and do not mind receiving goods from catalogues within a reasonable delivery time. The Operation Manager has focused on the utilisation of own resources to minimise operational costs.

Figure 6.5 shows the ratings of objectives, the actual performance and highlights the misalignment. It is evident that further examination is required for Timing and Material.

As shown in Figure 6.6, there is a conflict between Cost and Materials and further attention or a change of policy is required to resolve this conflict.

Training requirement

The training workshop for the RU/CS Analysis is likely to be of half a day's duration and it could be combined with the training for other tools. The understanding of rating and alignment could be simple, but the identification of conflicts is likely to require a few hands-on exercises.

	Machinery/Space	People	Materials
Utilisation objectives	3	3	1
Actual utilisation	3	3	2
Alignment	✓	✓	✗

	Specification	Cost	Timing
Customer service objectives	1	3	2
Actual level of service	2	3	1
Alignment	✗	✓	✗

✓ Good
✗ Issues to look at

1 = low, 2 = medium, 3 = high

Figure 6.5 The balance of objectives: mail order company.

3 High relative importance

Machinery/Space	People	Materials
3	3	1

1 Low relative importance

Specifications	Cost	Time
1	3	2

	Machinery/Space	People	Materials
Specifications			
Cost			✗
Time			

✗ Conflict

Figure 6.6 RU/CS conflicts in a mail order company.

Final thoughts

The RU/CS Analysis is a simple but powerful tool to establish quickly the conflicts between RU and CS and to reflect upon how to go about resolving them. This balance will vary between different operations or organisations.

A4: SWOT Analysis

Definition

An SWOT (Strengths, Weaknesses, Opportunities and Threats) is a tool for analysing an organisation's competitive position in relation to its competitors.

In the context of a quality improvement programme, an SWOT Analysis refers to a summary of the gaps and positive features of a process following the analytical stage.

Application

Based on an SWOT framework, the team members can focus on alternative strategies for improvement. For example, an S/O (Strengths/Opportunities) strategy may be considered to consolidate the strengths and open further leverage from the process. Similarly an S/T (Strengths/Threats) strategy may be considered to maximise the strength of the process and minimise risks.

Thus an SWOT Analysis can help the team to identify a wide range of alternative strategies for the next stage.

Basic steps

1. Create two categories (internal factors and external factors) and then further sub-divide into positive aspects (Strengths and Opportunities) and negative aspects (Weaknesses and Threats).
2. Ensure that the internal factors may be viewed as a strength or weakness of a process depending on their impact on the outcome or the process output.
3. Similarly assess the external factors bearing in mind that threats to one process could be an opportunity for another process.
4. Summarise the key features and findings derived from previous analyses in each of the SWOT categories.
5. Develop the improvement strategy for the next stage as pointers from the SWOT Analysis.

Worked-out example

The following example is taken from Kotabe and Helsen (2000), p. 277.

Table 6.1 shows the framework of an SWOT Analysis.

Table 6.1 The framework of an SWOT Analysers

External factors	Internal factors	Strengths	Weaknesses
		Brand name, human resources, technology, advertising	Price, lack of financial resources, long product development cycle
Opportunities	Growth market, de-regulation, stable exchange rate, investment grant	S/O strategy Maximise Strengths and maximise Opportunities	W/O strategy Minimise Weaknesses and maximise Opportunities
Threats	New entrants, change in consumer preference, local requirements	S/T strengths Maximise Strengths and minimise Threats	W/T strategy Minimise Weaknesses and minimise Threats

Training requirements

The training of the SWOT Analysis should be conducted in stages. The first part is in a classroom where the basic principles can be explained based on a known marketing product. The second part of the training is delivered to the team member when the results following the analysis of process are summarised.

Final thoughts

Although an SWOT Analysis is primarily a marketing tool, it can be applied effectively in order to summarise the key features of a process after the analytical stage.

A5: PESTLE Analysis

Definition

The PESTLE (Political, Economic, Social, Technical, Legal and Environmental) Analysis is an analytical tool for assessing the impact of external contexts on a project or a major operation and also the impact of a project on its external contexts. There are several possible contexts including:

- Political
- Economic
- Social
- Technical
- Environmental.

This is remembered easily by the acronym 'PESTLE' or 'Le Pest' in French. It is thus also known as the PEST Analysis.

Application

Very few major changes, whether they are caused by a major operation or a project, are unaffected by the external surrounds.

- *Political*: A project is affected by the policies of international, national or local government. It is also influenced by company policies and those of stakeholders, managers, employees and trade unions.
- *Economic*: The project is affected by national and international economic issues, inflation, interest rates and exchange rates.
- *Social*: The change is influenced by social issues, the local culture, the lives of employees, communications and language.
- *Technological*: The success of implementation is affected by the technology of the industry and the technical capability of the parent company.

- *Legal*: The project is affected by the legal aspects of planning, registration and working practices.
- *Environmental*: The impact of the change on environmental emission, noise, health and safety is assessed.

Basic steps

A PESTLE Analysis is carried out in four stages:

1. Develop a good understanding of the deliverables of the operation and the project. At this stage, the relevant policy and guidelines of both the local company and the parent organisation are reviewed.
2. List the relevant factors affecting the various aspects of the project related to PESTLE. It is important that the appropriate expertise of the organisation is drawn into the team for this analysis.
3. Validate the factors in Step 2 with the stakeholders and functional leaders of the company.
4. Review progress and decide on the next steps by asking two questions:
   a. How did we do?
   b. Where do we go next?
5. For more information on the PESTLE Analysis, see Turner et al. (2000, pp. 165–215).

Worked-out example

Consider the policy renewal management process of an insurance company based in Finland. The company implemented an on-line renewal process within Finland and wanted to expand the process in the European Union (EU).

A PESTLE Analysis was carried out as shown in Table 6.2.

The PESTLE Analysis shows that the expansion of on-line service, in general, has direct influence upon and opens up opportunities in the EU.

Training requirements

The principles of PESTLE Analysis are best learned by the process of the group working together during the project life cycle. The programme of Black Belt training normally includes a session on PESTLE Analysis and team leaders should receive a broad understanding of this tool. The analysis is of a strategic nature and thus may not involve all members of the project team.

Final thoughts

A PESTLE Analysis is most appropriate for the total programme rather than being used for individual operations. We recommend that this tool should be applied to all Six Sigma and Operational Excellence initiatives.

Table 6.2 PESTLE Analysis

Contexts	Key factors	Impact on company (0–10)
Political	• Within EU, countries are moving towards a more common political structure	6
Economic	• Slowing economy of Finland; GDP forecast to grow by 3.9% in 2003 • Dynamic changes in client business environment in Finland and Europe	8
Social	• Difference in buying habits in Finland versus EU	7
Technological	• Accelerating pace of change in ICT in Finland • On-line opportunity in EU with little extra cost • New cyber related risk in client business	9
Legal	• New constraints or requirements initiated by regulatory bodies (e.g. Insurance Supervisory Authority in EU)	9
Environmental	• No significant impact	1

A6: The Five Whys

Definition

The Five Whys is a systematic technique of asking five questions successively. The aim is to probe the causes of a problem and thus hopefully get to the heart of the problem.

Application

The Five Whys is a technique that is widely used to analyse problems in both manufacturing and service operations. It is a variation on the classic Work Study approach of 'critical examination', involving six questions: Why, What, Where, When Who and How?

The objective is to eliminate the root cause rather than patch up the effects.

Basic steps

1. Select the problem for analysis.
2. Ask five 'close' questions, one after another, starting with why.
3. Do not defend the answer or point the finger of blame at others.
4. Determine the root cause of the problem.

Worked-out example

The following example is taken from Stamatis (1999), p. 183.
Consider a problem: 'Deliveries are not completed by 4 p.m.'

Question 1: Why does it happen?
Answer: The routing of trucks is not optimised.

Question 2: Why is it not optimised?
Answer: Goods are loaded based on their size rather than the location of the delivery.

Question 3: Why are they loaded by size?
Answer: The computer defines the dispatch based upon the principle of 'large items first'.

Question 4: Why are large items given preference?
Answer: Large items are delivered first.

Question 5: But why?
Answer: Current prioritisation policy puts large items first on the delivery schedule.

Training requirements

As can be seen from the above question and answer model, the principle of this analytical tool is very straightforward. Thus the application of the Five Whys does not require any rigorous classroom training. The members of a problem solving team can easily understand and apply this simple tool after just one such group exercise.

Final thoughts

The Five Whys is an uncomplicated but very effective tool that can be used to identify the root causes of a problem. We recommend that, taking advantage of the fact that it is such a quick and unfussy approach, it can be applied on a far wider basis than at present.

A7: Interrelationship Diagram

Definition

An Interrelationship Diagram (ID) is an analytical tool to identify, systematically analyse and classify the cause and effect relationships among all critical issue of a process. The key drivers or outcomes are identified leading to an effective solution.

Application

An ID is often applied to enable the further examination of causes and effects after these are recorded in a Fishbone Diagram. ID encourages team members to think in multiple directions rather than merely in a linear sense.

This simple tool enables the team to set priorities to root causes even when credible data does not exist.

Basic steps

1. Assemble the team and agree on the issue or problem for investigation.
2. Lay out all of the ideas or issues that have been brought from other tools (such as a Cause and Effect Diagram) or brainstormed.
3. Look for the cause and effect relationships between all issues and assign the 'relationship strength' as:
   3 – Significant
   2 – Medium
   1 – Weak
4. Draw the final ID in a matrix format and insert the 'relationship strength' given by members.
5. Total the relationship strength in each row to identify the strongest effect of an issue on the greatest number of issues.

Worked-out example

The following example is taken from Bassard and Ritter (1994), p. 81.

Consider five key issues to improve CS.

1. Logistics Support
2. Customer Satisfaction
3. Education and Training
4. Personal Incentives
5. Leadership

The ID is plotted in a matrix (see Figure 6.7) with appropriate 'relationship strengths.'

From the above analysis in the 'Total' column, it is evident that Customer Satisfaction and Leadership are the two most critical issues.

Training requirements

The team can be adept in the application of ID after less than 1-hour's training in a classroom or a practical environment. The principles are simple, but it is important that a consensus is reached in attributing the relationship strength numbers.

	Logistics Support	Customer Satisfaction	Education and Training	Personal Incentives	Leadership	Total
Logistics Support		○	□	△	□	8
Customer Satisfaction	○		□	○	□	10
Education and Training	□	□		□	○	9
Personal Incentives	△	○	□		○	9
Leadership	□	□	○	○		10

Relationship strength: Key

○ Significant 3 □ Medium 2 △ Weak 1

Figure 6.7 Interrelationship Diagram.

Final thoughts

ID is not an essential tool to analyse the cause and effect relationship of issues, but because of its simplicity we recommend that it is used selectively to set priorities to causes or issues.

A8: Overall Equipment Effectiveness

Definition

The Overall Equipment Effectiveness (OEE) is an index of measuring the delivered performance of a plant or equipment based on good output.

The method of monitoring OEE is devised in such a way that it would highlight the losses and deficiencies incurred during the operation of the plant and identify the opportunities for improvement.

There are many ways to calculate OEE (see Shirose, 1992; Hartman, 1991). In this section we describe the methodology of OEE that was developed and applied by the author in both Unilever[1] and GlaxoWellcome.[2]

[1]In Unilever Plc, the methodology was known as PAMCO (Plant and Machine Control).
[2]In GlaxoWellcome it was called CAPRO (Capacity Analysis of Production).

Overall OEE is defined by the following formula:

$$OEE\% = \frac{\text{Actual Good Output}}{\text{Specified Output}} \times 100$$

where Specified Output = Specified Speed × Operation Time.

Application

The application of OEE has been extensive, especially when driven by the TPM (Total Productive Maintenance) programmes, to critical plant and equipment. It can be applied to a single equipment, a packing line, a production plant or processes. In order to appreciate the usefulness of OEE it is important to understand Equipment Time Analysis as shown in Figure 6.8 and described below.

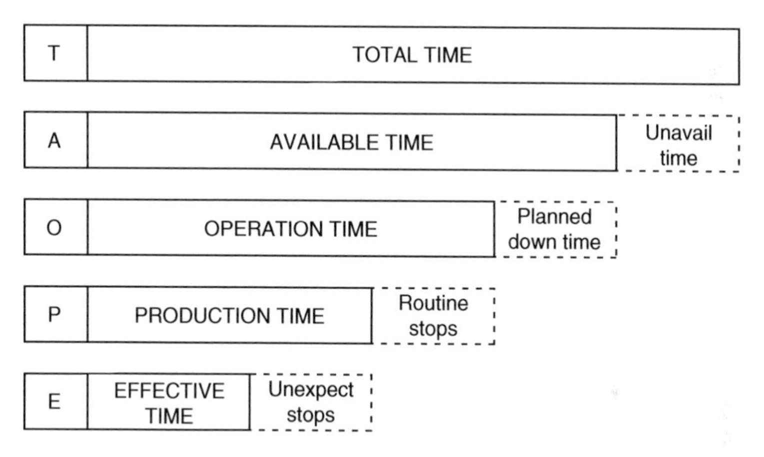

Figure 6.8 Equipment Time Analysis.

Total Time defines the maximum time within a reporting period, such as 52 weeks a year, 24 hours a day, 8760 hours in a year.

Available Time is the time during which the machine or equipment could be operated within the limits of national or local statutes, regulation or convention.

Operation Time is the time during which the machine or equipment is planned to run for production purposes. The operational time is normally the shift hours.

Production Time is the maximum time during which the machine or equipment could be expected to be operated productively after adjusting the Operation Time for routine stoppages such as changeover and meal breaks.

Effective Time is the time needed to produce a 'good output delivered' if the machine or equipment is working at its Specified Speed for a defined period. It includes no allowances for interruptions or any other time losses.

It is important to note that Effective Time is not recorded, it is calculated from the Specified Speed as

$$\text{Effective Time} = \frac{\text{Good Output}}{\text{Specified Speed}}$$

where Specified Speed is the optimum speed of a machine or equipment for a particular product without any allowances for loss of efficiency. It is expressed as quantity per unit such as Tons per hour, Bottles per minute, Cases per hour or Litres per minute.

In addition to OEE, two other indices are commonly used as shown below:

$$\text{Production Efficiency (\%)} = \frac{\text{Effective Time (E)}}{\text{Production Time (P)}} \times 100$$

$$\text{Operational Utilisation (\%)} = \frac{\text{Operation Time (O)}}{\text{Total Time (T)}} \times 100$$

A properly designed and administered OEE scheme offers a broad range of benefits and a comprehensive manufacturing performance system. Some of its key benefits are:

- It provides information for shortening lead time and changeover time and a foundation for SMED (single minute exchange of dies).
- It provides essential and reliable data for capacity planning and scheduling.
- It identifies the 'six big losses' of TPM leading to a sustainable improvement of plant reliability.
- It provides information for improving asset utilisation and thus reduced capital and depreciation costs in the longer term.

Basic steps

1. Select the machines, equipment or a production line where the OEE scheme could be applied. The selection criteria will depend on the criticality of the equipment in the context of the business. It is useful to start with a single production line as a trial or pilot.
2. Establish the Specified Speed of the production line governed by the control or bottleneck operation. As shown in the following example (Figure 6.9) of a soap packaging line, the Specified Speed is 150 tablets

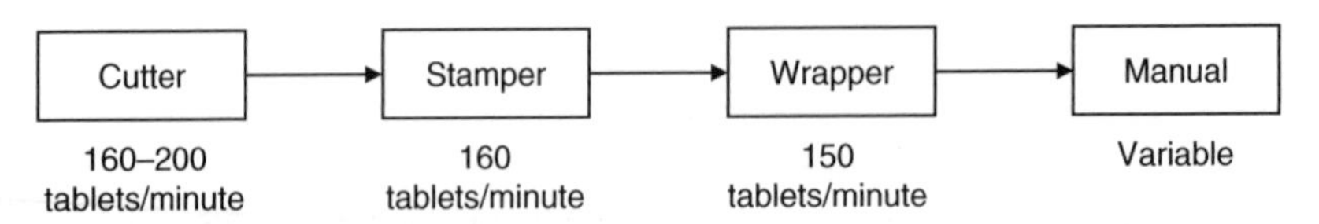

Figure 6.9 Soap production line.

per minute, i.e. this constitutes the speed of the wrapper (which is the slowest piece of equipment).

3. Set up a data recording system so that the output data and various stoppages and losses can be recorded.
4. Compile the data every day and validate the results. At this stage, detailed calculations are not necessary.
5. Monitor the results, comprising OEE and key indices, major losses as a per cent of the Operation Time and the trends of indices. The reporting is normally on a weekly basis for the department and on a monthly basis for Senior Management.
6. Use the results for continuous improvement, planning and strategic changes.

Worked-out example

Consider the production data of a toilet soap packing line where the control station governing the Specified Speed is an ACMA 711 wrapping machine.

Week Number:	31
Operation Time:	128 hours
Specified Speed:	150 tablets per minute
Good Output:	4232 cases
Routine Stoppages:	11 hours 30 minutes
Unexpected Stoppages:	27 hours 15 minutes

Given that each case contains 144 tablets

$$\text{Good Output} = 4232 \times 144 = 609\,408 \text{ tablets}$$

$$\text{Effective Time} = \frac{\text{Good Output}}{\text{Specified Speed}} = \frac{609\,408}{150 \times 60} = 67.71 \text{ hours}$$

$$\text{Production Time} = \text{Operation Time} - \text{Routine Stoppages}$$

$$= 128 - 11.5 = 116.5 \text{ hours}$$

$$\text{Total Time} = 7 \times 24 = 168 \text{ hours}$$

$$\text{OEE} = \frac{\text{Effective Time}}{\text{Production Time}} = \frac{67.71}{128} = 0.53 = 53\%$$

$$\text{Production Efficiency} = \frac{\text{Effective Time}}{\text{Production Time}} = \frac{67.71}{116.5} = 58\%$$

$$\text{Operation Utilisation} = \frac{\text{Operation Time}}{\text{Total Time}} = \frac{128}{168} = 76\%$$

It is important to note that the Effective Time was calculated and not derived from the recorded stoppages. There will be an amount of unrecorded time (also known as Time Adjustment) as, in the example, given by:

$$\begin{aligned}\text{Unrecorded Time} &= (\text{Production Time} - \text{Unexpected Stoppages}) \\ &\quad - \text{Effective Time} \\ &= (116.5 - 27.25) - 67.71 \\ &= 21.54 \text{ hours}\end{aligned}$$

Training requirements

The success of an OEE scheme depends heavily on the rigour of continuous training. It is important that each operator, supervisor and manager of a Production Department receives a half-day training programme covering the definitions, purpose and application of the OEE Scheme. The training is continuous because of the turnover of staff. Senior Management should also benefit from a 1-hour awareness session.

Final thoughts

The principles of OEE are conceptually simple but detail rich. The main strength of this tool is that it highlights the areas of deficiency for improvement and the key results cannot be manipulated. The specified time is calculated from tangible 'good output', Operation Time is well established shift hours and Total Time is absolute.

A9: TRIZ: Innovative Problem Solving

Definition

TRIZ is the Russian acronym for Teoriya Resheniya Izobreatatelskikh Zadatch (Inventive or Innovative Problem Solving). It extends traditional systems engineering approaches and provides powerful systematic methods for problem formulation, systems and failure analysis.

There are 39 characteristics and 40 principles of Innovative Problem Solving. The contradictions generated by 39 characteristics are systematically eliminated by 40 principles to lead to the development of anew inventions.

Application

There are two groups of problems people face: those with generally known solutions and those with unknown solutions. Those with known solutions can usually be solved by information found in books, technical journals or with subject matter experts. The other type of problem is one with no known solution. It is called an inventive problem and may contain contradictory requirements. Methods such as brainstorming and trial-and-error are commonly suggested. A better approach was developed by Genrich S. Altshuller (1994), born in the former Soviet Union in 1926 and this approach later known as TRIZ should satisfy the following conditions:

1. Be a systematic, step-by-step procedure
2. Be a guide through a broad solution space to direct to the ideal solution
3. Be repeatable and reliable and not dependent on psychological tools
4. Be able to access the body of inventive knowledge
5. Be able to add to the body of inventive knowledge
6. Be familiar enough to inventors by following the general approach to problem solving in Figure 6.10.

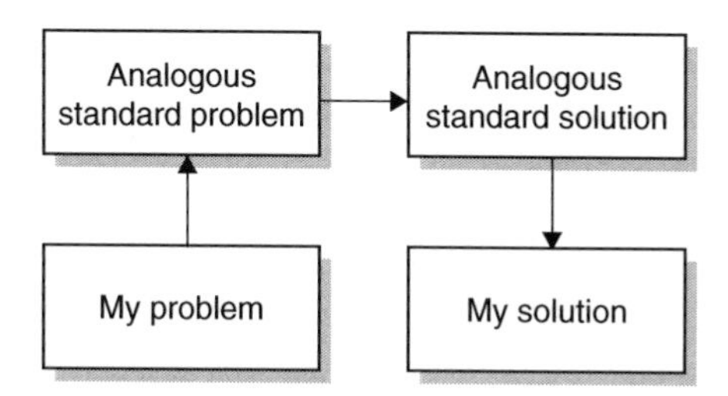

Figure 6.10 TRIZ general problem solving model.

Altshuller more clearly defined an inventive problem as one in which the solution causes another problem to appear (such as increasing the strength of a metal plate causing its weight to get heavier) and categorised the solutions into five levels.

- *Level one*: Routine design problems solved by methods well known within the specialty. No invention needed. About 32% of the solutions fell into this level.
- *Level two*: Minor improvements to an existing system, by methods known within the industry. Usually with some compromise. About 45% of the solutions fell into this level.
- *Level three*: Fundamental improvement to an existing system, by methods known outside the industry. Contradictions resolved. About 18% of the solutions fell into this category.
- *Level four*: A new generation that uses a new principle to perform the primary functions of the system. Solution found more in science than in technology. About 4% of the solutions fell into this category.
- *Level five*: A rare scientific discovery or pioneering invention of essentially a new system. About 1% of the solutions fell into this category.

TRIZ has been used in Russia to solve technical problems and develop thousands of patentable inventions. It is now gaining acceptance in major companies in Western World and Pacific Rim and also in Six Sigma projects. Samsung and Intel Corporation have recently reported successful application of TRIZ in innovation and projects.

Basic steps

There are four basic steps of TRIZ as shown in Figure 6.10. These steps which follow a theme of increasing identity are:

Step 1. Identifying my problem – Identify the engineering system being studied, its operating environment, resource requirements, primary useful function, harmful effects and ideal result.

Step 2. Formulate the problem the Prism of TRIZ – Restate the problem in terms of physical contradictions and identify problems that could occur. For example, identify if by improving one technical characteristic to solve a problem cause other technical characteristics to worsen, resulting in secondary problems arising

Step 3. Search for previously well-solved problem – Find the contradicting engineering characteristics (from 39 TRIZ technical characteristics). First find the characteristic that needs to be changed. Then find the characteristic that is an undesirable secondary effect. State the standard technical conflict.

Step 4. Look for analogous solutions and adapt to my solution – Altshuller also extracted from the worldwide patents 40 inventive principles and the Table of Contradictions. These are hints that will help an engineer find a highly inventive (and patentable) solution to the problem. The Table of Contradictions lists the 39 Engineering Parameters on the x-axis (undesired secondary effect) and y-axis (feature to improve). In the intersecting cells, are listed the appropriate Inventive Principles to use for a solution.

Worked-out examples

A well known example of TRIZ is the redesign of a metal beverage can. The first step is to identify the problem goal called the Initial Final Result (IFR). The IFRs of our example include the design of a cylindrical metal container to hold the beverage of a given volume that can support the weight of stacking to human height without causing damage to cans or the contents held by the container.

The standard technical conflicts are as follows. If we make the containers thinner more stress will be felt by the container walls. On the other hand if we make the walls of the containers thicker, the containers would be heavier and the stacking of heavy containers could damage the bottom layer. Furthermore heavy containers would be more expensive. This conflict of increasing strength and increasing weight could be reconciled by Segmentation, one of the 40 TRIZ principles. The answer is to change the smooth cylindrical surface of the container to a wavy surface made up of many little walls. The new design will increase the edge strength of the cylinder while retaining its lighter weight.

Training requirements

TRIZ is a highly specialised problem solving tool. Although it is conceptually simple comprising four basic steps it very rich in details. My consulting companies are guarding the 39 technical characteristics and 40 principles along with the apparently complex Contradiction matrix to promote training workshops. The training workshops are usually covered 3–5 days. During the course the participants learn about problem identification and formulation and search for solutions by applying the matrix of 39 characteristics and 40 principles.

Final thoughts

TRIZ is conceptually a very powerful problem solving tool leading to new ideas and innovative solutions. The bottom line question is 'if TRIZ is so so good, why isn't everyone using it?' The answer is that because of its prescriptive approach with the baggage of 39 characteristics and 40 principles it is difficult to sell and it also competes with other DFSS tools such as QFD.

7

Tools for improvement

Invention, strictly speaking, is little more than a new combination of those images which have been previously gathered and deposited in the memory; nothing can come of nothing.

— Joshua Reynolds, circa 1780

The earlier stages of the project, in particular the Analyse stage, have pinpointed the areas for improvement. During the Improve phase, the ideas and solutions are put to work. Various options are then compared with each other to determine the most promising solution. Some experiments and trials may be required to validate the best solution. Finally, this solution is usually piloted on a small scale in the business environment.

The objectives of the tools of the Improve phase are to help the team to develop a solution for improving process performance and to confirm that the proposed solution will meet or exceed the quality improvement goals of the project.

The key deliverables of the Improve phase are:

1. *Proposed solution*: A solution for reducing variation or eliminating the special causes of the problem in the process.
2. *Validate solution*: Process improvement that has been piloted in a real business environment.

The important tools for improvement should include:

I1: Affinity Diagram
I2: Nominal Group Technique
I3: SMED
I4: Five S
I5: Mistake Proofing
I6: Value Stream Mapping
I7: Brainstorming
I8: Mind Mapping
I9: Force Field Diagram

The Improvement stage also depends on techniques including Brainstorming, Design of Experiments (DOE), Quality Function Deployment (QFD) and Failure Mode and Effect Analysis (FMEA).

I1: Affinity Diagram

Definition

An Affinity Diagram is an improvement tool to generate creatively a number of ideas and then summarise logical groupings among them to understand the problem and then to lead to a solution.

This is also known as the KJ method, identified with Kawakita Jiro, the Japanese scientist who first applied it in the 1950s.

Application

An Affinity Diagram is used to categorise verbal information into an organised visual pattern. It is used in conjunction with brainstorming when problems are uncertain, large and complex, thereby enabling the user to create a discipline out of chaos.

It is a tool to overcome 'team paralysis' which is created by the generation of a large number of options and a lack of consensus. As part of brainstorming all ideas are recorded on sticky notes or index cards. The ideas on the notes are then clustered into major categories. It aims to be a creative, rather than a logical process. Hence the Affinity Diagram is regarded more as an improvement tool rather than an analytical one.

Basic steps

1. Clarify the chosen problem or opportunity in a full sentence.
2. Collect the current data available on the problem or opportunity. This is usually done by brainstorming within a group.
3. Record each piece of data or idea onto a card or Post-It™ note and place them at random onto a board.
4. Sort ideas simultaneously into related groups.
5. Arrange the group affinity cards, usually less than 10 in each group, in a logical order.
6. For each grouping, label header cards and draw broad lines around the group affinity cards.

Worked-out example

The following example is adapted from Schmidt et al. (1999, p. 125).

Consider the problem as clearly stated below:

What are the barriers to a quality improvement programme?

Figure 7.1 shows an example of an Affinity Diagram related to the above problem.

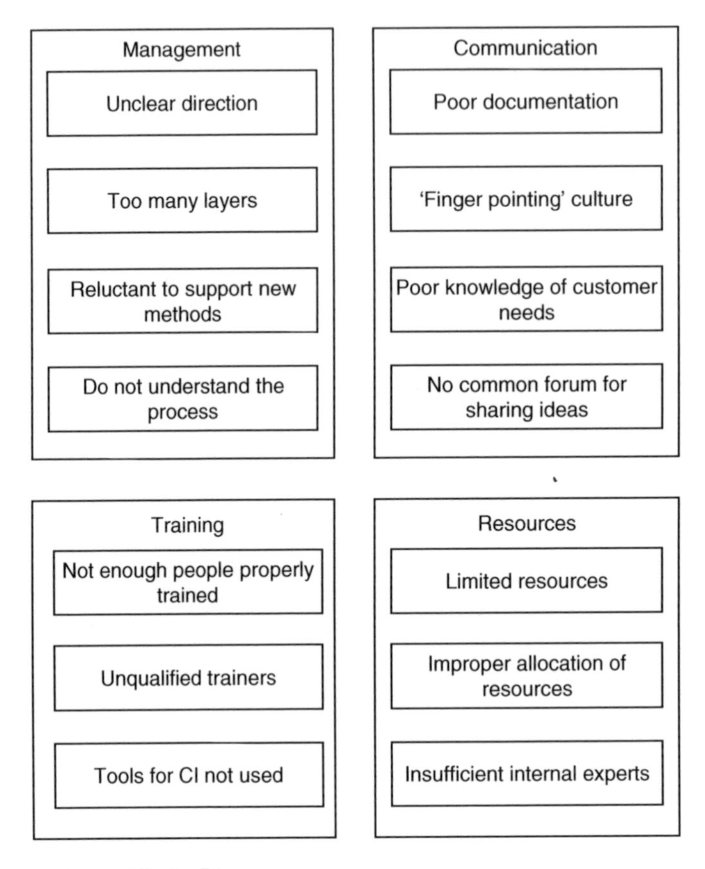

Figure 7.1 An Affinity Diagram.

Training requirements

The knowledge and application of an Affinity Diagram are best derived by 'on-the-job' training during a brainstorming exercise on an actual problem. The facilitator should have experience in applying a number of Affinity Diagrams to direct the team and gain a consensus of grouping.

Final thoughts

It is important to note that an Affinity Diagram is an improvement tool used in conjunction with a brainstorming exercise. It should not be confused with CEDAC, a hybrid of the Cause and Effect Diagram, which also uses cards or Post-It™ notes.

I2: Nominal Group Technique

Definition

Nominal Group Technique (NGT) is an improvement tool to derive an importance ranking in a team's list of ideas arising from a brainstorming exercise.

This is also known as the Weighted Multi-voting Technique.

Application

NGT is widely used to arrive quickly at a consensus on the relative importance of ideas or issues in a group working environment. It allows every team member to rank issues without any pressure from others. Thus the quieter individuals exercise the same power as the more dominant team members.

The application of NGT usually follows a brainstorming session. Incidentally, you may have recognised a similar process being applied during the voting procedure for the Eurovision Song Contest! (Now you have a good reason for watching it next time.)

Basic steps

1. Assemble the team and generate a list of issues, problems or ideas to be prioritised. This could be done either by a brainstorming session or each member may be asked to write down their ideas.
2. Write the list on a flip chart or board. Refine the list by eliminating duplicate or similar statements.
3. Assign labels (A, B, C, etc.) to the final list of statements and record them on a flip chart or board.
4. Each member records the labels of each statement in a rank order with the highest number allocated to the most important statement. For example, if there are five statements then '5' is the most important and '1' is the least important ranking.
5. Aggregate the rankings of all team members and the statement with the highest point would have the highest priority.

Worked-out example

Consider the case of a private school where several complaints have been made about the low morale of staff, but they are of an informal and unstructured nature and the Board of Governors have little to work with. Table 7.1 shows the team's list of complaints.

Five members of the team have allocated a rank order to each problem listed in Table 7.1 and the results are shown in Table 7.2.

Training requirements

Similar to the 'Affinity Diagram', the knowledge and application of NGT can be best derived by 'on-the-job' training in an actual group exercise. The facilitator should have the experience of directing a team and gaining a consensus.

Table 7.1 List of problems

A	Not enough teamwork and consultation
B	Information coming from an 'in crowd'
C	No consistent procedures
D	Dogmatic approach of the Head
E	Lack of feedback from reports made to parents and pupils
F	Strong hierarchy and work demarcation
G	Lack of training for support staff
H	More focus on fees than education
I	Lack of credit or recognition for support staff
J	Inadequate workspace except for the Head and 'cronies'

Table 7.2 Nominal Group Technique

Problem	Team members					Total	Priority
	1	2	3	4	5		
A	7	9	9	6	6	37	
B	10	7	7	9	8	41	High
C	6	10	8	6	7	37	
D	9	8	10	8	10	45	High
E	5	3	6	7	9	30	
F	8	5	4	5	3	25	
G	3	4	5	4	2	18	
H	1	2	1	3	1	8	
I	4	5	2	2	5	18	
J	2	1	3	1	4	11	

Final thoughts

NGT is not a 'scientific' approach but it is simple and builds commitment to the team's choice through equal participation. It generates an atmosphere of fun and individual inclusion, building 'team spirit'. We recommend the application of NGT for solving in particular problems related to cultural and 'softer' issues.

I3: SMED

SMED or Single Minute Exchange of Dies is name of the approach used for reducing output and quality losses due to changeovers and setups.

'Single Minute' means that necessary setup time is counted on a single digit.

Application

This method has been developed in Japan by Shigeo Shingo (1985) and has proven its effectiveness in many manufacturing operations by reducing the changeover times of packaging machines from hours to minutes.

The primary application area of SMED is the reduction of setup times in production lines. This process enables operators to analyse and find out themselves why the changeovers take so long and how this time can be reduced. In many cases, changeover and setup times can be condensed to less than 10 minutes, so that the changeover time can be expressed with one single digit, and it is therefore called 'SMED'.

SMED is considered as an essential tool in Lean Manufacturing and it is instrumental in the reduction of non-value added activities in process times. Changeover loss is one of the six big losses that have been defined within the Total Productive Maintenance (TPM). It is important to note that SMED is directly linked with the analytical process of OEE (Overall Equipment Effectiveness).

With due respect to the success of the SMED method, it is fair to point out that the basic principles are fundamentally the application of classical industrial engineering or work study.

Basic steps

1. Study and measure the operations of the production line to discriminate:
   - Internal setup, the operation that must be done while machine is stopped.
   - External setup, the operation that possibly can be done while the machine is still running.
2. Suppress non-value added operations and convert internal setup operating into external setup. The data from OEE and the preparations of prerequisites (e.g. tools, changeover parts, pre-assemblies, pre-heating, mobile storage, etc.) are reviewed to achieve results. Some internal setups are converted to external setups.
3. The next stage is to simplify the design of the machine, especially fillings and tightening mechanisms. Some examples of the design simplification are U-shaped washers, quarter turn screws and cam and lever tights.
4. Balance the work content of the line and ensure teamwork. For example, in one automatic insertion machine, one operator sets up on the machine front while the other operator feeds components on the other side.
5. Minimise trials and controls. Use of Mistake Proofing or Poka-Yoke enables the standard way to be carried out each time.

Worked-out example

The following example is taken from Basu and Wright (1997, p. 97).

Consider the setup time reduction of a packing machine.

The internal and external setup times have been measured. As shown in Figure 7.2, the total setup time is reduced by overlapping external setup times on the internal setup time.

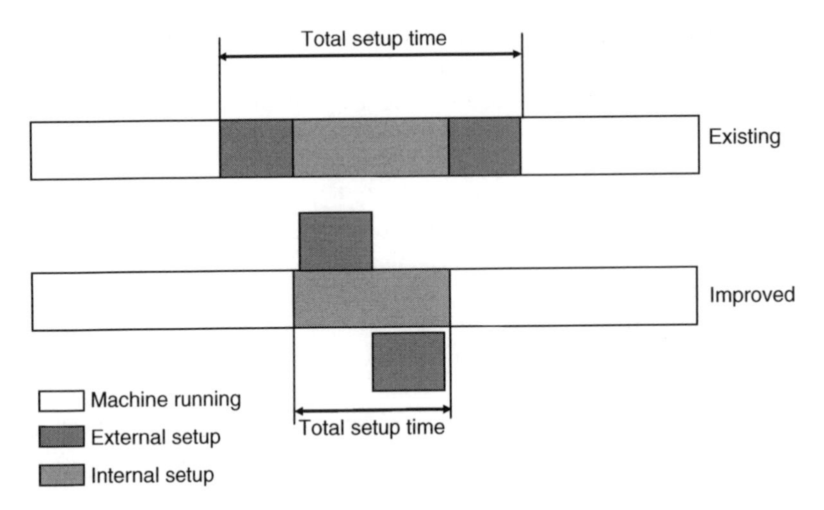

Figure 7.2 SMED: Setup time reduction.
(*Source*: Basu and Wright, 1997).

Training requirements

The training of SMED is likely to be more effective on team members with a good understanding of work study or industrial engineering. The basic principles should be explained to members with hands-on exercises being carried out in a room. This should be supplemented by a production line study in a factory environment. It is also important that the basics of OEE are covered during an SMED training session.

Final thoughts

SMED can produce excellent results in achieving the reduction of setup times but admittedly it is resource intensive. It should therefore be restricted to bottleneck operations.

I4: Five S

Definition

Five S is a tool for improving the housekeeping of an operation, developed in Japan, where the five Ss represent five Japanese words all beginning with 's':

- *Seiri (Organisation)*: Separate what is essential from what is not.
- *Seiton (Neatness)*: Sort and arrange the required items in an orderly manner and in a clearly marked space.
- *Seiso (Cleaning)*: Keep the workstation and the surrounding area clean and tidy.

- *Seiketson (Standardisation)*: Clean the equipment according to laid down standards.
- *Shitsuke (Discipline)*: Follow the established procedure.

In order to retain the name 'Five S', a number of English language versions have evolved. These include:

- Seiri: Sort
- Seitor: Set in order/Stabilise
- Seiso: Shine
- Seiketsu: Standardise
- Shitsuki: Sustain.

Application

The Five S method is a structured sequential programme to improve workplace organisation and standardisation. Five S improves the safety, efficiency and the orderliness of the process and establishes a sense of ownership within the team.

Five S is used in organisations engaged in Lean Sigma, Just-in-Time (JIT), Total Productive Maintenance (TPM) and Total Quality Management (TQM). This principle is widely applicable not just for the shop floor, but for the office too. As an additional bonus there are benefits to be found in environmental and safety factors due to the resulting reduced clutter. Quality is improved by better organisation and productivity is increased due to the decreased time spent in searching for the right tool or material at the workstation. Consider the basic principle of a parent tidying a small child's room which is overflowing with clutter and sorting together various types of toys. The end product should be a neater, warmer, brighter and more civilised play environment which will encourage the child to utilise all toys and equipment more productively because all relevant pieces are together, space is enhanced and mess is reduced.

It is useful to note that the quality gurus of Japan like numbered lists, e.g. the Seven Mudas, the Five Whys, the Five Ss. However, the exact number of Ss is less important than observing the simple doctrine of achieving the elimination of wastes.

As the Five S programme focuses on attaining visual order and visual control, it is a key component of Visual Factory Management.

Basic steps

1. *Sort*: The initial step in the Five S programme is to eliminate excess materials and equipment lying around in the workplace. These non-essential items are clearly identified by 'red-tagging'.
2. *Set in order*: The second step is to organise, arrange and identify useful items in a work area to ensure their effective retrieval. The storage area, cabinets and shelves are all labelled properly. The objective of this step is, as the old mantra says, 'a place for everything and everything in its place'.

3. *Shine*: This third action point is sometimes known as 'sweep' or 'scrub'. It includes down-to-basics activities such as painting equipment after cleaning, painting walls and floors in bright colours and carrying out a regular cleaning programme.
4. *Standardise*: The fourth point encourages workers to simplify and standardise the process to ensure that the first three steps continue to be effective. Some of the related activities include establishing cleaning procedures, colour coding containers, assigning responsibilities and using posters.
5. *Sustain*: The fifth step is to make Five S a way of life. Spreading the message and enhancing the practice naturally involves people and cultural issues. The key activities leading to the success of Five S include:
   - Recognise and reward the effort of members
   - Top management awareness and support
   - Publicise the benefits.
6. The final step is to continue training and maintaining the standards of Five S.

Worked-out examples

As Five S is primarily a visual process, a good example of promoting its message would be to display pictures of a workplace with photographs showing both 'before' and 'after' depictions of the implementation of Five S.

The following example is taken from Skinner (2001) to illustrate the benefits of a Five S programme.

Northtrop Grumman Inc. in the United States first deployed Five S on a part delivery process. The work area assembled a variety of components into a single product.

Before Five S, the area was not well organised, and the process was inefficient. With Five S implementations, the area saw a huge 93% reduction in the space employees travel to complete tasks as well as a 42% reduction in the overall floor space.

The system has become a one-piece flow operation between assembly and mechanics, enabling everyone involved to know what the station has and what it needs.

Training requirements

Five S is a conceptually simple process, but it requires both initial and follow-up training to inculcate the methodology to all employees. The classroom training sessions should be followed by, as far as practicable, a visit to a site where visual changes due to Five S could be observed. A second best option is to show the members photography or videos illustrating the 'before and after' status of the workplace involved in a Five S programme.

Final thoughts

Five S is a simple tool and should be considered for the housekeeping and visual control of all types of work area, whether they are in manufacturing or service.

I5: Mistake Proofing

Definition

Mistake Proofing is an improvement tool to prevent errors being converted into defects. It comprises two main activities: preventing the occurrence of a defect and detecting the defect itself.

Mistake Proofing is also known as Poka-Yoke. The concept was developed by Shigeo Shingo and the term 'poka-yoke' comes from the Japanese words 'poka' (inadvertent mistake) and 'yoke' (prevent).

Application

Mistake Proofing is applied in fundamental areas. Although Poka-Yoke was devised as a component of Shingo's 'Zero Quality Control' for Toyota production lines, it is very easy to understand and grounded in basic common sense.

The process of Mistake Proofing is simply paying careful attention to every activity in the process and then placing appropriate checks at each step of the process. Mistake Proofing emphasises the detection and correction of mistakes at the Design stage before they become defects. This is then followed by checking. It is achieved by 100% inspection while the work is in progress by the operator and not by the quality inspectors. This inspection is an integral part of the work process.

There are abundance of examples of simple devices related to Mistake Proofing in our everyday surroundings including limit switches, colour coding of cables, error detection alarms, a level crossing gate and many more.

Basic steps

1. Perform Shingo's 'source inspection' at the Design stage. In other words, identify possible errors that might occur in spite of preventive actions. For example, there may be some limit switches that provide some degree of regulatory control to stop the machine automatically.
2. Ensure 100% inspection by the operator to detect that an error is either taking place or is imminent.
3. Provide immediate feedback for corrective action. There are three basic actions in order of preference:
   i. *Control*: An action that self-corrects the error, e.g. spell checker.
   ii. *Shutdown*: A device that shuts down the process when an error occurs, e.g. a limit switch.
   iii. *Warning*: Alerts the operator that some error is imminent, e.g. alarm.

Worked-out example

Consider the situation leading to the development of a level crossing. This is a place where cars and trains are crossing paths and the chances of accidents are very high.

The possible errors that might occur would relate to car drivers, who might be thinking one thing or another or distracted while driving (source inspection).

Both the level crossing operator and the car driver should ensure safety features while the work is in progress (judgement inspection).

In order to prevent drivers from making mistakes when a train is approaching, traffic lights were installed to alert the driver to stop (warning).

The lights might not be completely effective, so a gate was installed when a train was coming (shutdown or regulatory function).

The operation of the gate was controlled automatically as the train was approaching (control).

With the above Mistake Proofing devices in place, an accident can only occur if either the control and regulatory measures are malfunctioning or the driver drives around the gate.

Training requirements

There is no 'rocket science' involved in Mistake Proofing and it may be perceived in a dismissive fashion: 'that's only common sense'. However, it is critical that there should be some basic training in the principles and applications of Mistake Proofing. Furthermore, the employees need to be empowered to make improvements in the process by using Mistake Proofing. A half-day workshop should meet these training requirements.

Final thoughts

Mistake Proofing is a simple tool in principle, but its execution is the difficult part. The contrast between Mistake Proofing and 'fool-proofing' however is critical. The essential difference is that in Mistake Proofing, operators are respected and treated as partners in solving problems.

I6: Value Stream Mapping

Definition

Value Stream Mapping (VSM) is a visual illustration of all activities required to bring a product through the main flow, from raw material to the stage of reaching the customer.

Mapping out the activities in a production process with cycle times, down times, in-process inventory and information flow paths helps us to visualise the current state of the process and guides us to the future improved state.

Application

VSM is an essential tool of Lean Manufacturing in identifying non-value added activities at a high level of the total process.

According to Womack and Jones (1996), the initial objective of creating a Value Stream Map is to identify every action required to make a specific product. Thus the initial step is to group these activities into three categories:

1. Those which actually create value for the customer.
2. Those which do not create value but are currently necessary (type one Muda).
3. Those which create no value as perceived by the customer (type two Muda).

Once the third set has been eliminated, attention is focused on the remaining non-value creating activities. This is achieved through making the value flow at the pull of the customer.

VSM is closely linked with the analytical tool of Process Mapping. Having established improvement opportunities at a high level by VSM, a detailed analysis of the specific areas of the process is effective with Process Mapping.

Basic steps

1. The first step of VSM is to select the product or process for improvement.
2. Each component of production from the source to the point of delivery is then identified.
3. The entire supply chain of the product or process (e.g. through order entry, purchasing, manufacturing, packaging and shipping) is mapped sequentially.
4. The quantitative data of each activity (e.g. storage time, delay, distance travelled, process time and process rate) are then recorded.
5. Each component (i.e. activity) of production or process is evaluated to determine the extent to which it adds value to product quality and production efficiency.
6. These activities are then categorised as:
   - Value added
   - Necessary non-value added
   - Unnecessary non-value added.
7. Areas of further analysis and improvement are then identified clearly.

Worked-out example

The following example is adapted from Womack and Jones (1996, pp. 38–43).

Consider a case containing eight cans of cola at a Tesco store.

Figure 7.3 shows a Value Stream Map of cola, from the mining of Bauxite (the source of aluminium of the cans) to the user's home.

The quantitative data related to the activities in the value stream are summarised in Table 7.3.

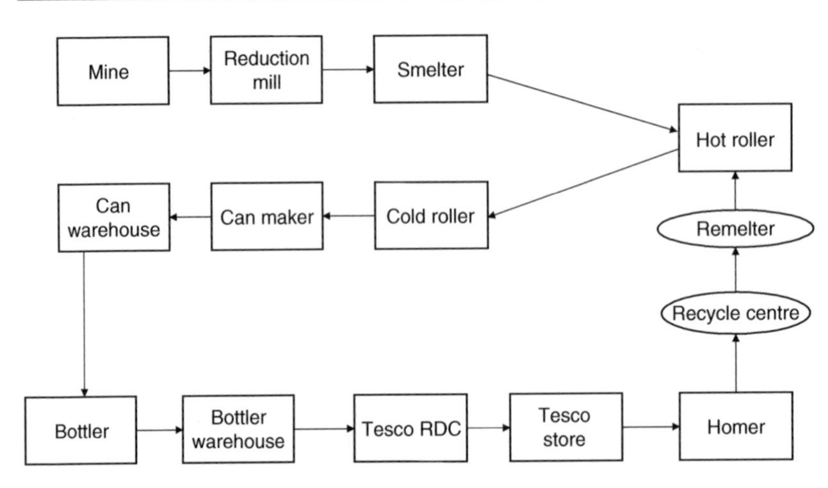

Figure 7.3 Value Stream for Cola Cans (© Ron Basu).

Table 7.3 Quantitative data of Cola Cans

	Incoming storage	Process time	Finished storage	Process rate	Cumulative days
Mine	0	20 minutes	2 weeks	1000 tonnes/ hour	319
Reduction mill	2 weeks	30 minutes	2 weeks	–	305
Smelter	3 months	2 hours	2 weeks	–	277
Hot rolling mill	2 weeks	1 minute	4 weeks	10 feet/ minute	173
Cold rolling mill	2 weeks	<1 minute	4 weeks	2100 feet/ minute	131
Can maker	2 weeks	1 minute	4 weeks	2000 feet/ minute	89
Bottler	4 days	1 minute	5 weeks	1500 feet/ minute	47
Tesco RDC	0	0	3 days	–	8
Tesco store	0	0	2 days	–	5
Home storage	3 days	5 minutes	–	–	3
Totals	5 months	3 hours	6 months	–	319

It is evident from the details in Table 7.3 that value added activities take only 3 hours compared to the total time (319 days) from the mine to the recycling bin. This proportion is surprisingly small when one considers the overall duration of the process.

Training requirements

The basic principles of VSM are not new, but making sense of these ideas and applying them to practical problems clearly require some training. There is no shortage of training consultants offering workshops and courses in the tools of Lean Manufacturing including VSM.

We recommend that the team members should undergo a training workshop of at least half a day's duration for VSM. This training programme should be combined with other relevant tools like Process Mapping.

Final thoughts

A complete Value Stream Map quickly provides the visibility of the total process and is very effective in identifying non-value added activities at a macro level.

Summary

Arguably the most difficult and certainly the most creative part of the Six Sigma and Operational Excellence initiatives is the Improvement stage. It is not rational to expect that the improvement tools described in this section will point out the obvious way forward. The solutions depend on the knowledge, innovative ideas and teamwork of the members. The tools and techniques are there to channel the ideas and analytical data towards improvement.

I7: Brainstorming

Definition

Brainstorming is an improvement tool for a team to generate, creatively and efficiently, a high volume of ideas on any topic by encouraging free thinking.

There are a few variations on the brainstorming process, of which two methods are more frequently used. First is the structured method (known as the 'round robin') where each member is asked to put forward an idea. The other technique is unstructured and is known as 'free-wheeling', in which ideas are produced and expressed by anyone at any time.

Application

Brainstorming is employed when the solution to a problem cannot be found by quantitative or logical tools. It works best by stimulating the synergy of a group. One member's thoughts trigger the idea of another participant, and so on. It is often used as a first step to open up ideas and explore options, and these are then followed up by appropriate quality management tools and techniques.

It has the advantage of getting every member involved, avoiding a possible scenario where just a few people dominate the whole group.

There are some simple ground rules or codes of conduct to observe:

- Agree to a time limit with the group
- Accept all ideas as given and do not interpret or abbreviate
- Do not evaluate ideas during the brainstorming process.
- Encourage quantity rather than quality of ideas.
- Discourage the role of an expert
- Keep ideas expressed in just a few words
- Emphasise causes and symptoms as opposed to solutions
- Write clearly and ensure the ideas are visible to everyone
- Have fun!

Basic steps

1. Clearly state the focused problem selected for the brainstorming session.
2. Form a group and select a facilitator, agree on a time limit and remind members of the ground rules.
3. Decide whether a structured approach or a free-wheeling basis will be used. For a larger group, a structured approach will allow everyone to get a turn and subsequently this could be switched to the free-wheeling method.
4. Write clearly on a flip chart or a board any ideas as they are suggested. The facilitator will motivate and encourage participants by prompting them, 'What else?'
5. Review the clarity of the written list of ideas, allow them to settle and discard any duplication.
6. Apply filters to reduce the list. Typical filters could include cost, quality, time and risk.
7. Ensure that everyone concurs with the shortlist of ideas.

Worked-out example

Consider the following focused statement for brainstorming.

What are the key selection criteria of a family holiday?

The five members of a family generated 26 ideas or issues. These were then filtered by a budgeted cost of £4000 for the whole family and the following key criteria were derived:

- Two weeks in August
- Seaside resort
- Indoor and outdoor recreational facilities
- Not near a nightclub
- Rich local culture
- Opportunities for sightseeing.

Training requirements

The application of brainstorming does not require any formal training in a classroom. A facilitator with some previous experience in the process can conduct a successful brainstorming session after briefing the team with the ground rules.

Final thoughts

Brainstorming is a very useful tool for generating ideas in a group. Follow the ground rules with particular emphasis on two points:

1. Do not dominate the group
2. Set a time limit of, say, half an hour for the entire session.

I8: Mind Mapping

Definition

Mind Mapping is a learning tool for ordering and structuring the thinking process of an individual or team working on a focused theme.

According to Buzan (1995), the Mind Map 'harnesses the full range of cortical skills – word, image, number, logic, rhythm, colour and spatial awareness – in a single, uniquely powerful technique'.

It is a graphic tool to express 'radiant thinking' comprising four key characteristics:

1. The subject or theme is presented as a central image or key word.
2. The main components of the subject root out from the central image as branches.
3. Each branch contains a key word printed on an associated line.
4. The sub-components of each branch are also represented as branches attached to higher level branches.

Mind Map is arguably comparable to the Cause and Effect Diagram where the effect represents the central image of the Mind Map. Each of the branches of the Mind Map are the causes in the Cause and Effect Diagram.

Application

Mind Maps have been applied for both individual and group objectives.

The applications in individual areas included:

- Note taking
- Multi-dimensional memory device
- Creative thinking.

Buzan (1995) claims that the mode of note taking involved in making a Mind Map saves more than 90% of the total time required for the conventional linear method of making notes. Instead, the Mind Map version involves noting, reading and reviewing only relevant key words.

Another way of thinking, the mnemonic device, involves the use of the imagination and association in order to produce multi-dimensional memorable images. The use of the memory Mind Map activates the brain to become mnemonically alert and thus increases the memory skill level.

The Mind Map is suited to creative thinking because it uses all the cortical skills commonly associated with creativity, especially imagination, association of ideas and flexibility.

The Group Mind Map becomes a powerful tool during a group brainstorming exercise. The Mind Map becomes a 'hard copy' of the emerging group consensus and at the same time reflects the evolution of ideas through the branches and sub-branches radiating from the central image or theme.

In recent years, Group Mind Maps have been successfully used by universities (including Oxford and Cambridge) and multi-national companies like Boeing Aircraft Corporation, EDS, Digital Computers and British Petroleum.

Basic steps

The Mind Map is intended to increase mental freedom and thus it is important not to introduce rigid disciplines. However there is a need for a structured approach, otherwise freedom may be mistaken for chaos.

Buzan (1995) suggests six 'Mind Map laws' and offers three recommendations to supplement these 'laws'. The 'laws' are:

1. Use a central image and emphasise that image by using variations in size of printing and colour.
2. Use arrows when you want to make connections within and across branch patterns.
3. Be clear and use only one key word per line.
4. Develop a personal style, while maintaining the Mind Map 'laws'.
5. Use hierarchy and categorisation in the form of basic ordering of ideas.
6. Use a numerical order simply by numbering the branches in the desired order.

The three recommendations are designed to help you implement the 'laws'. The recommendations are:

a. Break mental blocks
   i. Add blank lines to your ongoing Mind Map
   ii. Ask questions to stimulate a block-breaking response
   iii. Add images to your Mind Map
   iv. Maintain awareness of your associational capacity.
b. Reinforce your Mind Map
   i. Review your Mind Maps
   ii. Do quick Mind Map checks.

c. Prepare
   i. Your mental attitude
   ii. Your materials
   iii. Your workplace/environments.

The above 'laws' and recommendations are applicable to both individual and Group Mind Maps. It is important to designate a facilitator to process a Group Mind Map starting with a central image.

Worked-out example

An example of a Mind Map for Late Delivery is illustrated in Figure 7.4.

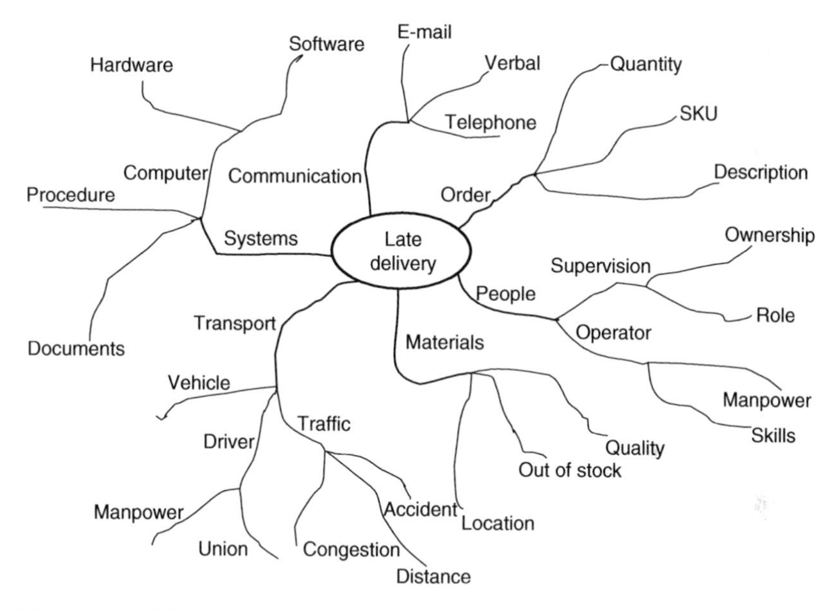

Figure 7.4 A Mind Map (© Ron Basu).

In this Mind Map, Late Delivery is the central image. The main causes are noted by key words on radial lines and sub-causes are shown on branches emerging from these radial lines.

If after the completion of the Mind Map, the importance of root causes can be weighted by assigning a number 1–100 according to its importance.

Training requirements

The basics of Mind Mapping can be grasped by reading Tony Buzan's 'The Mind Map Book' (1995). However it would be useful if a facilitator with previous experience in Mind Mapping conducts a number of trial exercises before applying it to a real life problem.

Final thoughts

Mind Mapping has various individual and group applications. It is particularly useful as a mnemonic or analytical tool for developing personal choices. It gives the brain a wider range of information on which to base its decision.

19. Force Field Analysis Diagram

Definition

The Force Field Analysis Diagram or simply Force Field Diagram is a model built on the concept by Kurt Lewin (1951) that change is characterised by as a state of equilibrium between driving forces (e.g. new technology) and opposing or restraining forces (e.g. fear of failure).

In order for any change to happen the driving forces must exceed the restraining forces, thus shifting the equilibrium.

Application

Force Field Diagram is a useful tool at the early stage of change management leading to improvement. It is often used to

- investigate the balance of power involved in an issue or obstacle at any level (personnel, project, organisation, network);
- identify important players or stakeholders – both allies and opponents;
- identify possible causes and solutions to the problem.

Basic steps

According to Lewin (1951) three key steps are involved in the concept of change management by Force Field Diagram:

1. First, an organisation has to unfreeze the driving and restraining forces that hold it in a state of apparent equilibrium.
2. Second, an imbalance is introduced to the forces, by increasing the drivers or reducing the restrainers or both, to enable the change to take place.
3. Third, once the change is complete and stable, the forces are brought back to equilibrium and refrozen.

Force Field Diagram is constructed by a team and the following basic steps are suggested:

1. Agree on the current problem or issue under investigation and the desired situation.
2. List all forces driving changes towards the desired situation.
3. List all forces resisting changes towards the desired situation.

4. Review all forces and validate their importance.
5. Allocate score to each of the forces using a numeric scale (e.g. 5 = most important and 1 = least important).
6. Chart the forces by listing the driving forces on the left and restraining forces to the right.
7. Decide how to minimise or eliminate the restraining forces and increase the driving force.
8. Agree on an action plan.

Worked-out example

The issue identified is how to increase the usage of purchase orders in a pharmaceutical company. The driving and restraining forces were identified by the team and also rated in a scale 1–5 (1 = low, 5 = high) and represented in a Force Field Diagram as shown in Figure 7.5.

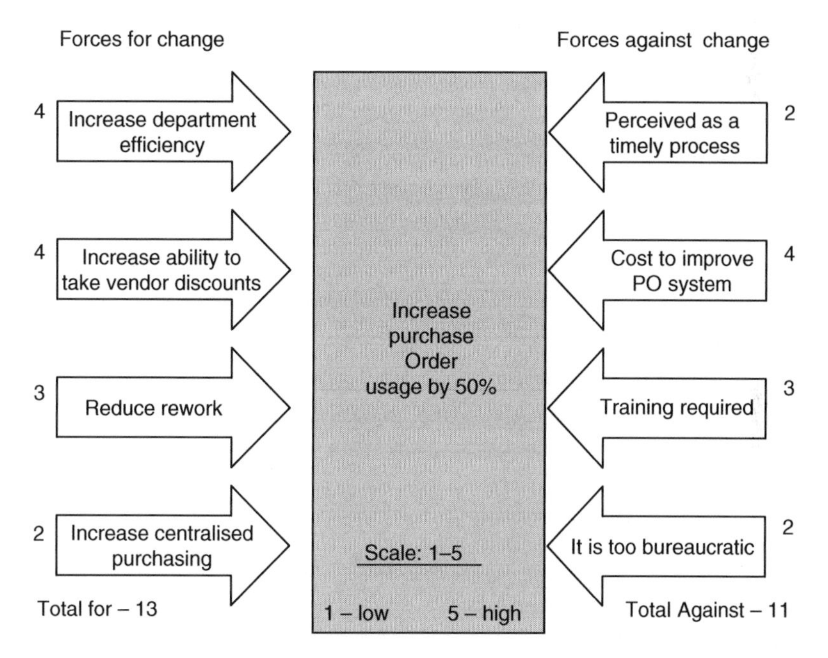

Figure 7.5 Force Field Analysis Diagram.

The driving forces show a total score of 13 against a total score of 11 shown by the restraining forces.

Training requirements

The knowledge and application of a Force Field Diagram are best derived by 'on-the-job' training during a brainstorming exercise on an actual problem.

The facilitator should have experience change management in applying a number of Force Field Diagrams to direct the team and gain a consensus of identifying forces and allocating scores.

Final thoughts

Force Field Diagram is a powerful tool in the Improvement stage of Six Sigma and also in tempering the mindset in any change management projects. It could also provide new insights into the evaluation and implementation of corporate strategies.

8

Tools for control

That is what learning is. You suddenly understand something you have understood all your life, but in a new way.

— Doris Lessing, 'The Four-Gated City'

Introduction

Operational Excellence is a long-term process. It can take an organisation several years to put the fundamental principles, practices and systems into action in a learning culture that will sustain the benefits gained. Managers who have steered their way through the challenges of a Six Sigma or a Lean Sigma programme over recent years are probably proud of the results. However, such a manager has only just embarked upon the path of success.

The objective of the 'Control' phase is to implement the solution, ensure that this solution is sustained, and share the lessons learned from the improvement projects throughout the organisation. With this approach the projects start to create excellent returns. Thus the best practices in one part of the organisation are translated quickly to result in implementation in projects carried out by another part of the organisation.

The main deliverables of the Control stage are:

1. Project documentation – a close-out report to record the key aspects of the project.
2. Leverage of best practice – transfer of key learning from your project that may be adopted in other projects.
3. Sustained solution – a fully implemented process that is supported with a Control plan to ensure that it is sustained over time.

The key tools for Control should include:

C1: Gantt Chart
C2: Activity Network Diagram
C3: Radar Chart

C4: PDCA Cycle
C5: Milestone Tracker Diagram
C6: Earned Value Management

Some of the tools from the early stages can be used during the implementation of the project (e.g. M7: Control Charts). In order to ensure the sustainability of the results, a number of qualitative techniques (see Chapter 10) are applicable at the Control stage. These include:

- Balanced Scorecard
- European Foundation of Quality Management
- Sales and Operations Planning.

C1: Gantt Chart

Definition

A Gantt Chart is a simple tool which represents time as a bar or a line on a chart. The start and finish times for activities are displayed by the length of the bar and often the actual progress of the task is also indicated.

A Gantt Chart is also known as a Bar Chart.

Application

The most common form of scheduling is the application of Gantt Charts. The merits of Gantt Charts are that they are simple to use and they provide a clear visual representation of both the scheduled and actual progress of activities. The current time is also indicated on the graph.

Gantt Charts are also used to review alternative schedules by using movable pieces of paper or plastic channels. The charts can be drawn easily by standard software tools such as PowerPoint or Excel. However, a Gantt Chart is not an optimising tool and therefore does not determine the 'critical path' of a project.

Basic steps

1. Identify the key activities or the tasks related to the project and describe each activity by selective key words.
2. Prepare a scheduling board, and depending on the duration of the project, draw vertical lines to divide the board in monthly, weekly or daily intervals.
3. Arrange the activities in a sequence of estimated start dates and post them on the extreme left-hand column of the board.
4. Estimate the start and finish dates of each activity and draw horizontal bars or lines along the time scale to reflect the start and duration of each of the activities.

5. On completion of each activity, show the actual start and duration of the activity by a bar of a different colour.
6. Include a 'Time Now' marker on the chart; review and maintain the chart until the end of the project.

Worked-out example

Figure 8.1 shows an example of a Gantt Chart showing the planned and completed activities of a FIT SIGMA™ programme.

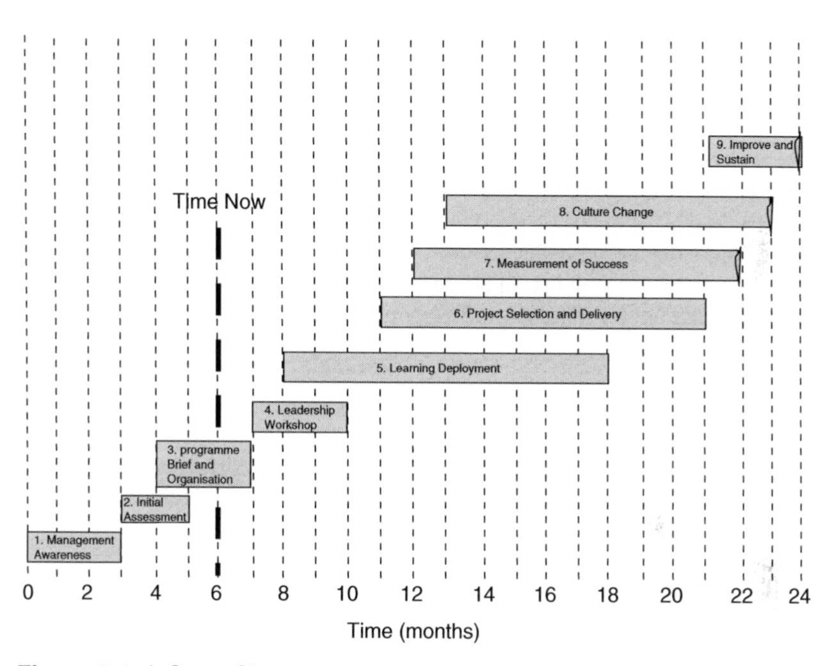

Figure 8.1 A Gantt Chart.
Source: Basu and Wright, 2003, p. 170.

Training requirements

The application of Gantt Charts does not require extensive training. The team members are usually experienced in the use of Gantt Charts. A briefing session in front of the scheduling board should be adequate.

Final thoughts

A Gantt Chart is a simple but very effective visual tool for planning and monitoring the progress of a quality programme. We recommend its application at the Control stage of the programme.

C2: Activity Network Diagram

Definition

An Activity Network Diagram is a control tool to determine and monitor the most efficient path, known as the critical path, and a realistic schedule for the completion of a project. The diagram is represented graphically showing a brief description of all tasks, their sequence, their expected completion time and the jobs that can be carried out simultaneously.

An Activity Network Diagram with some variations is also referred to as PERT (Project Evaluation and Review Technique), CPM (Critical Path Method), a Precedence Diagram and finally, as Network Analysis.

Application

The Activity Network Diagram was extensively used in most projects during the 1960s and 1970s. As the larger projects became more and more complex comprising numerous tasks, its popularity by manual methods started to diminish. However, with the advent of software systems such as Primavera and MS Project, its application at the higher level of the project has increased significantly. It offers a number of benefits:

- The team members can visualise the criticality of major tasks in the overall success of the project.
- It highlights the problems of 'bottlenecks' and unrealistic timetables.
- It provides facilities to review and adjust both the resources and schedules for specific tasks.

Basic steps

There are normally two methods applied for the construction of an Activity Network Diagram: the 'activity on arrow' method and the 'activity on node' method. It is the former that has been used most widely and the steps for the application of the arrow technique can be outlined as follows:

1. Assemble the project team with the ownership and knowledge of key tasks.
2. List the key tasks with a brief description for each one.
3. Identify the first task that must be done, the tasks that can be done in parallel and the sequential relationship between tasks.
4. Draw arrows for each task which are labelled between numbered nodes and estimate a realistic time for the completion of each of these tasks.
5. Avoid feedback loops in the diagram. Unlike Gantt Charts, the length of the arrows does not have any significance.
6. Determine the longest cumulative path as the critical path of the project.
7. Review the Activity Network Diagram and adjust resources and schedules if appropriate.

For more detailed information on the Activity Network Diagram, see Wild (2002, pp. 403–450).

Worked-out examples

Consider a project of writing and submitting the draft manuscript of a technical book such as this one to a publisher. Table 8.1 lists all the activities which constitute the project including their dependent relationship and estimated duration.

Table 8.1 List of project activities for production of a technical book

Resource	Activity	Description	Predecessor	Duration (weeks)
Author	A	Prepare proposal	–	2
Publisher	B	Approve proposal	A	4
Author	C	Preliminary research	A	2
Author	D	Detailed research	C	10
Author	E	Write Chapters 1–3	B	3
Author	F	Write Chapters 4–6	E	3
Author	G	Write remaining chapters	F	4
Admin	H	Type Chapters 1–3	E	2
Admin	I	Type Chapters 4–6	F, H	2
Admin	J	Type remaining	G, I	3
Author	K	Compile full draft	D, J	2
Author	L	Obtain copyright clearance	B	12
Author	M	Submit manuscript	K, L	1

The Activity Network Diagram is shown in Figure 8.2.

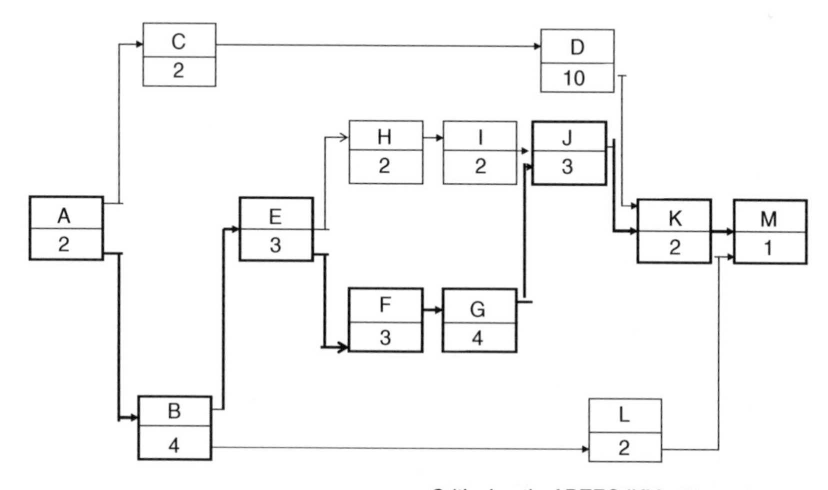

Figure 8.2 Activity Network Diagram (© Ron Basu).

The critical path is A, B, E, F, G, J, K and M with a total project duration of 22 weeks.

Training requirements

The basic principles of an Activity Network Diagram are not difficult to follow and can be covered in a half-day workshop with practical exercises.

A detailed analysis of an Activity Network Diagram can be complex when the variable estimates of duration, 'float' or 'slack', resource levelling and the probability of occurrence are considered. We suggest that the advanced applications should be treated for academic research.

Final thoughts

With the flexibility, simulation and milestone facilities now available on a software for Gantt Charts, we recommend that an Activity Network Diagram should be used primarily for the high level mapping of tasks to identify the critical path.

C3: Radar Chart

Definition

A Radar Chart is a polar graph to show using just one graphic the size of the gaps in the performance levels of key performance indicators.

A Radar Chart is also known as a Polar Graph and, because of its appearance, as a Spider Diagram.

Application

A Radar Chart is a useful visual tool to display the important metrics of performance at the Control stage of a quality improvement programme. The other benefits of this chart include:

- It highlights strengths and weaknesses of the total process, programme or organisation.
- It can define full performance in each category.
- It can act as a focal point to capture and review the different perception of all stakeholders of the organisation related to relevant performance metrics.
- Given a range of rating (say on a scale of 1–5), it can drive a total or average score of all entities.

However, a limitation of a Radar Chart is that it tends to provide just a snapshot of the performance levels at any given time.

Basic steps

1. Select and define the performance categories. The chart can handle 10–20 categories.
2. Some performance metrics are likely to be easily quantifiable and expressed as a percentage. Other metrics may be qualitative and not represented by numbers.
3. Normalise all performance metrics in a scale of 1–5 with appropriate guidelines according to the following grades:
   - Poor
   - Fair
   - Good
   - Very Good
   - Excellent.
4. Construct the chart by drawing a wheel with as many spokes as the performance categories and marking each spoke with '0' at the centre and '5' on the rim.
5. Connect the ratings for each performance category and highlight the gaps.
6. Use the results for consolidating strengths and improving weaknesses.

Worked-out example

The example of a Radar Chart is taken from Basu and Wright (1997, p. 199).

Consider that all the operations of a manufacturing organisation can be expressed in six categories and 20 metrics as given below:

Category (pillars)	Metrics (foundation stones)
Marketing and innovation	3
Environment and safety	3
Supply chain management	3
Manufacturing facilities	5
Procedures	3
People	3

Figure 8.3 shows a Radar Chart depicting a snapshot of 20 metrics.

Training requirements

The application of a Radar Chart does not require any significant classroom based training. However it does necessitate a good understanding of how to rate all performance categories. This understanding and the selection criteria of the metrics should be explained in a period of about an hour as part of a training programme.

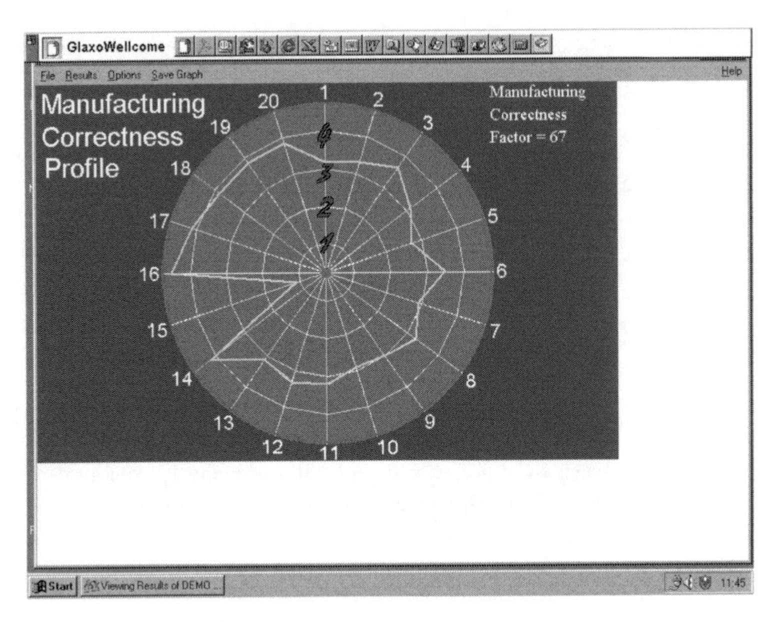

Figure 8.3 A Radar Diagram.
Source: Basu and Wright, 1997.

Final thoughts

Given the availability of appropriate data, a Radar Chart can be drawn quite simply using easily accessible software tools. We recommend its application to monitor the progress of a programme by a snapshot of all key performance indicators.

C4: PDCA Cycle

Definition

In a central process, the actual results of an action are compared with a target or a set point. The difference between the two is then monitored and corrective measures are adopted if the disparity becomes large. The repeated and continuous nature of continuous improvement follows this usual definition of Control and is represented by the PDCA (Plan, Do, Check, Act) Cycle.

This is also referred to as the Deming Wheel, named after W.E. Deming. Another variation of PDCA is PDSA (Plan, Do, Study, Act).

Application

The application of the PDCA Cycle has been found to be more effective than adopting the 'right first time' approach of concentrating on developing flawless plans (Juran, 1999, p. 41.3). The PDCA Cycle means continuously looking for better methods of improvement.

The PDCA Cycle is effective in both doing a job and managing a programme. The extent to which the PDCA Cycle is applied to the job level depends on the self-control of the operators. Education and training enhance the self-control capacity of workers. At the programme level, the PDCA Cycle acts as a process of repeatedly questioning the detailed working of the operations and thereby helps to sustain the improved results.

The PDCA Cycle enables two types of corrective action – temporary and permanent. The temporary action is aimed at results by practically tackling and fixing the problem. The permanent corrective action, on the other hand, consists of investigating and eliminating the root causes and thus targets the sustainability of the improved process.

Basic steps

1. P (Plan) stage: The cycle starts with the Plan stage, comprising the formulation of a plan of action based on the analysis of the collected data.
2. D (Do) stage: The next step is the Do or implementation stage. This may involve a mini-PDCA cycle until the issues of implementation are resolved.
3. C (Check) stage: The next step is the Check stage where the results after implementation are compared with targets to assess if the expected performance improvement has been achieved.
4. A (Act) stage: At the Act stage, if the change has been successful then the outcome is consolidated or standardised.
5. If the change has not been successful, however, the lessons are recorded and the cycle starts again. Even if the change is successful, the results are sustained by going through the PDCA Cycle over and over again.

Worked-out example

The following example is taken from Juran (1999, pp. 32.8–32.10).

In this example, a Healthcare organisation in the United States was following the traditional reliance on extensive internal and external inspection to maintain quality standards. This resulted in a medical record system whose size, complexity and format were wasteful and cumbersome.

The aspects of the PDCA Cycle were applied to their internal quality assurance procedures and the medical record procedures were simplified.

Figure 8.4 shows the PDCA Cycle.

Training requirements

The importance of the training for the PDCA Cycle, especially for first time workers, has been well recognised at the early stage of the quality movement and thus the concept of the Quality Circle was born. During the late 1950s,

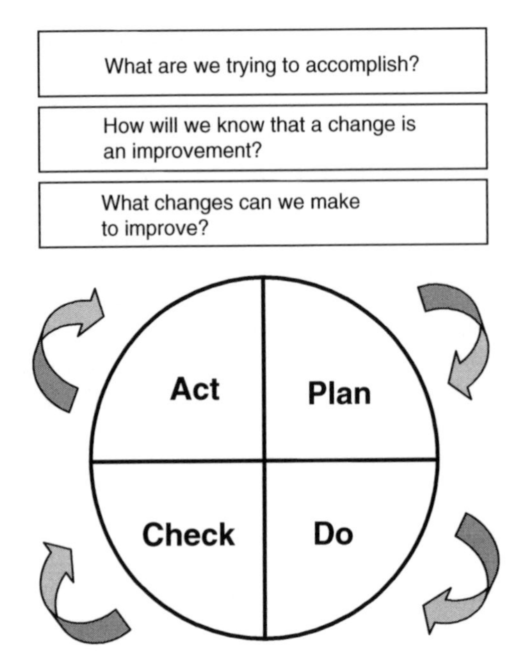

Figure 8.4 PDCA Cycle (© Ron Basu).

about 100 000 transcripts of radio broadcast text for Quality Circles were sold in Japan.

The training of the PDCA Cycle, although simple in principle, should be inculcated in everyone within the organisation on a continuous basis. This tool then becomes most effective when it becomes a way of life in the organisation.

Final thoughts

The PDCA Cycle is more than just a tool; it is a concept of continuous improvement processes embedded in the organisation's culture. The most important aspect of PDCA lies in the 'Act' stage after the completion of a project when the cycle starts again for further improvement.

C5: Milestone Tracker Diagram

Definition

A milestone is a key event selected for its importance in the project. A Milestone Tracker Diagram is used to show the projected milestone dates and the best estimated date on the week of the progress review on a single chart.

The purpose is to assess the achievement or slippage of the progress. Thus the Milestone Tracker Diagram is also called the Milestone Slippage Chart.

Application

A Milestone Tracker Diagram is used in a large and long running project to review the probability of achieving the key dates of the project. However, its principles can be applied to projects of shorter dimension.

The manager of the project or programme faces a problem when team members report their work as being 90% or 95% complete, leaving the last 10% or 5% continuously out of reach. The first requirement is to identify a set of key events or milestones in the plan that allows no compromise. A milestone is achieved when the objectives or targets of the key event are delivered in full. The dates of these milestones should not be far apart. A simple rule is that when the target passes from person to person or from department to department, the event must constitute a milestone. However the workplace of a person or department may also contain several milestones.

Basic steps

1. Determine the key events of the project as milestones.
2. Establish the planned delivery dates of the milestones and agree the dates with the Project Manager and Sponsor.
3. Draw two axes at right angles to each other and mark these axes in a suitable scale of weeks or months. The total period should allow the milestones to extend under the worst possible conditions.
4. Choose the x-axis as the 'Estimated Finish Week' of milestones and y-axis as the 'Week of Progress Review'. The axes should be of equal value. Note that the y-axis is in the reverse direction as compared to a conventional graph.
5. Draw a diagonal line from the top of the y-axis (zero) to the maximum duration point on the x-axis.
6. During every week of progress review, plot the estimated delivery dates of each milestone.
7. At any time during these progress reviews, the Project Manager can draw the 'best fit' straight line to connect these revised estimates. If these straight lines are projected they will intersect the diagonal line at points which can be taken as the predicted delivery dates of milestones.

Worked-out examples

Figure 8.5 shows an example of a Milestone Tracker Diagram where the progress of two milestones have been illustrated.

The planned finish dates of Milestone 1 and Milestone 2 are Week 4 and Week 8, respectively.

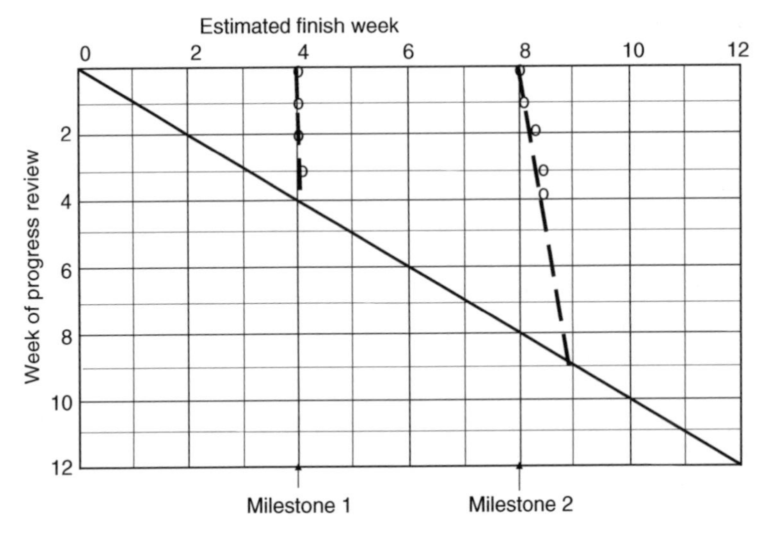

Figure 8.5 Milestone Tracker Diagram (© Ron Basu).

The diagram (Figure 8.5) shows that Milestone 1 is on track while Milestone 2 has experienced some delays and thus its predicted finish date is Week 9.

Training requirements

The basic principles of a Milestone Tracker Diagram are easy to follow and may not require any specific training in classrooms. The definition and guidance provided in this section of the book should enable the project team to apply this tool in practice.

Final thoughts

The Milestone Tracker Diagram is a useful, though not essential, tool for reviewing the progress of key events in a project. It is a better name than the 'Milestone Slip Chart', something of a misnomer because it implies that milestones are expected to slip!

C6: Earned Value Management

Definition

Earned Value Management or Earned Value Analysis is a project control tool for comparing the achieved value of work in progress against the project schedule and budget. It can be performed at the single activity level and by aggregating the results up through the hierarchy or work breakdown structure.

There are a few useful terms related to Earned Value Management as shown in Figure 8.6 and defined below:

Time Now: The reference point used to measure and evaluate the current status.

Earned Value: The value of useful work done at Time Now. It is also known as the Budgeted Cost of Work Performed (BCWP). It is typically calculated at activity level by multiplying the Budget at Completion (BAC) for the activity with the % progress achieved for the activity.

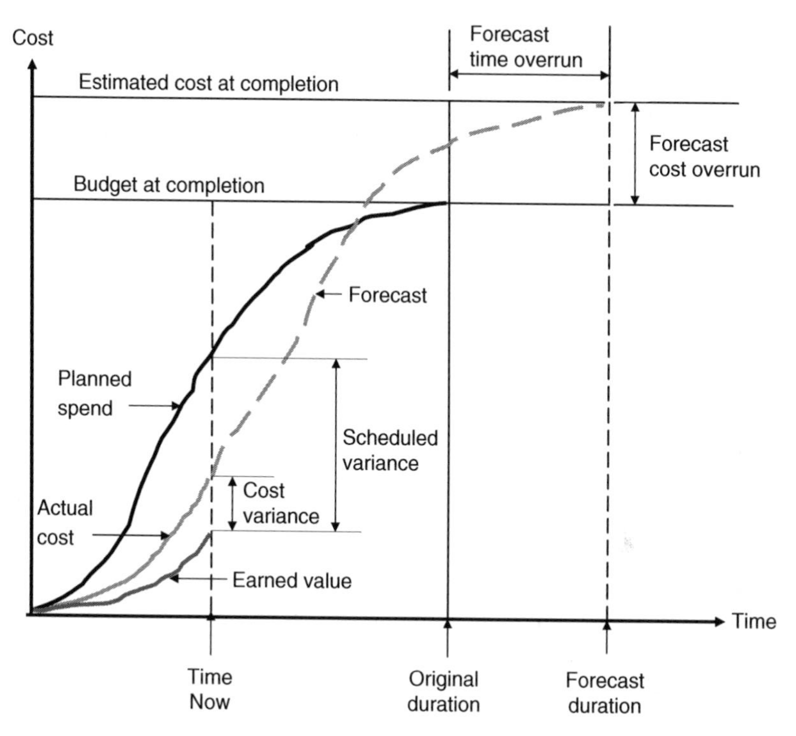

Figure 8.6 Earned Value Management (© Ron Basu).

Budget at Completion: The total budget for the work to be carried out.

Original Duration: The planned overall duration of the activity or project.

Planned Value (Spend): The planned rate of spend against time though the life of the project. It is also known as the Budget Cost of Work Schedule (BCWS).

Actual Cost: The cumulative costs incurred at Time Now. It is also known as the Actual Cost of Work Performed (ACWP).

Cost Variance (CV): The difference between the Earned Value and the Actual Cost at Time Now (CV = Earned Value − Actual Cost).

Scheduled Variance (SV): The difference between the Earned Value and the Planned Value at Time Now (SV = Earned Value − Planned Value).

Scheduled Performance Index (SPI): The ratio between the Earned Value and Planned Value at Time Now (SPI = Earned Value/Planned Value). It is used to predict the final outcome of the project time.

Cost Performance Index (CPI): The ratio between the Earned Value and Actual Cost at Time Now (CPI = Earned Value/Actual Cost). It is used to forecast the Estimated Cost at Completion (EAC).

$$EAC = \text{Actual Cost}/CPI + (BAC - \text{Earned Value})/CPI$$

Application

Earned Value Management has been applied effectively in well-structured projects, especially engineering projects, where there is a defined work breakdown structure and various levels of hierarchy.

It is not a progress control tool in itself, as it can only highlight a need for corrective action by indicating trends. At each milestone review (i.e. Time Now), it provides some useful pointers of project control including:

Variance analysis: It shows current status in terms of cost and schedule.

Estimating accuracy: It enables predictions of cost at completion and completion date.

Efficiency: It provides performance indices identifying areas requiring corrective action.

However, there is a danger of placing too much reliance on Earned Value Management, because the results are likely to be flawed for a number of reasons. These include:

- Forecasts depend on reliable measurements of the amount of work performed that can be difficult to achieve in some cost types.
- A considerable amount of clerical effort is needed to maintain the database and carry out calculations.
- Difficulty is likely to be greatest for projects containing a higher proportion of procured equipment and materials.
- It does not take into account risks and uncertainties.
- If the methodology of EVM is not fully understood by the sponsor and members of the project, it becomes difficult to get everyone's co-operation.

Basic steps

1. Establish a sound framework of planning and control for implementing Earned Value Management. The key features of this framework include:
   a. A detailed work breakdown structure
   b. A detailed cost coding system to match the work breakdown structure
   c. Timely and accurate collection of cost data

   d. A sound method of quantifying the amount of work done at each review date
   e. A well-trained administration.
2. At each review date, determine for the work package:
   a. Authorised current BAC
   b. % Work complete (ACWP)
   c. Planned spend (BCWS).
3. From the data in Step 3, calculate:
   a. Earned Value = Current Budget × % Complete = BCWP
   b. CV = BCWP − ACNP
   c. SV = BCWP − BCWS
   d. CPI = BCWP/ACWP
   e. SPI = BCWP/BCWS.
4. From the data in Step 4, calculate:
   a. Estimated Cost at Completion (EAC) = BAC/CPI
   b. Estimated Time at Completion = Original Duration/SPI.
5. Analyse the results and trends and take corrective action.

Worked-out example

Consider the example of a small project which has a BAC of £80 000 and duration of 6 weeks. The collected data at the end of Week 3 ('Time Now') are shown in Table 8.2.

Table 8.2

Project weeks	Planned value (£)	Actual cost (£)	% Complete
1	10	8	10
2	20	18	15
3	35	30	20
4	60		
5	70		
6	80		

Earned value at Week 3 (Time Now) = £80 × 0.20 = £16

Cost Variance = Earned Value − Actual Cost = 16 − 30 = −14

Schedule Variance = Earned Value − Planned Value = 16 − 35 = −19

Scheduled Performance Index = Earned Value/Planned Value
= 16/35 = 0.46

Cost Performance Index = Earned Value/Actual Cost = 16/30 = 0.53

Forecast Duration = Original Duration/SPI = 6 × 35 = 13 weeks

$$\text{Estimated Cost at Completion} = \text{BAC/CPI} = \frac{£80 \times 30}{16} = £150\text{k}$$

The project is currently overspending and in addition has fallen behind schedule.

Training requirements

The principles of EVM are conceptually simple but rich in detail. In the worked-out example, we have looked at only one activity but the Earned Value and its related parameters would have to include all project work scheduled to be completed at 'Time Now'. The Earned Value should include all work actually finished as well as the completed portion of all work in progress. Therefore it is important that the Administrator is trained thoroughly in a classroom environment, say in a full-day workshop. The members of the project team will also benefit from a half-day appreciation workshop.

Final thoughts

The detailed requirements of EVM should not deter the reader from its merits for implementation. We recommend that EVM should be applied with appropriate training in larger.

Summary of Part 2

The tools for managing quality have been grouped according to their most appropriate or frequently used area in the DMAIC cycle. This is not a rigid taxonomy; the purpose of this classification is primarily for presenting the tools in a systematic way. The same tool (e.g. Cause and Effect Diagram) can be used in Define, Measure, Analyse or Improve stages of the programme. The tools are not enough to make an improvement; they need adequate training, resources, management support and an environment which is conducive to change and to their application. A tool is as good as a person who trained to use it the right way for the appropriate problem.

Part 2: Questions and exercises

1. In your organisation, which processes have the highest priority for improvement?
   Which tools would you use to come to this conclusion?
2. In your local hospital, even after their admission to a specific ward, patients have to wait for their beds. Use the Cause and Effect Diagram to show the possible causes for the delay in assigning beds.
3. You are going to buy 10 different items from your local supermarket. The major activities involved would include parking the car, collecting your trolley, picking items you wish to buy, paying at the check-out, returning to your car and finally unloading purchases into it.
   Develop a Flow Process Chart by taking into account detailed elements of operations and identifying the non-value added activities.
4. What are the distinctive features of
   - Value Stream Mapping
   - Process Mapping
   - Flow Diagram
   - Flow Process Chart.

   Give examples of the areas of application of these tools. Explain how these tools can be applied for both service and manufacturing operations.
5. Discuss the relationship between
   a. Brainstorming, Nominal Group Technique and Mind Mapping
   b. A Run Chart, Scatter Diagram and Regression Analysis
   c. An Interrelationship Diagram and Affinity Diagram
   d. Control Charts and Process Capability Measurement
   e. Overall Equipment Effectiveness (OEE) and SMED
   f. The Five Ss, the Five Whys and Mistake Proofing
   g. A Gantt Chart, Activity Network Diagram and Milestone Tracker Diagram.
6. A travelling salesman has recorded the petrol consumption rate (miles per gallon or MPG) for 20 weeks as follows:

Week	1	2	3	4	5	6	7	8	9	10	11	12	13	14	15	16	17	18	19	20
MPG	21	19	20	22	18	21	20	20	18	20	19	22	20	21	19	20	19	21	18	22

   a. Draw a histogram of the above data
   b. Calculate the mean (μ) and standard deviation (σ) of the sample of MPG data
   c. Draw a Control Chart of the data showing both the UCL (Upper Control Limit) and LCL (Lower Control Limit) where $UCL = \mu + 3\sigma$ and $LCL = \mu - 3\sigma$.

   (see Appendix 3 to compare answers)

7. A toothpaste packaging line contains two machines, Arencomatic and Nordanmatic. Arencomatic is a twin-nozzle filler. Each nozzle can fill at a

maximum rate of 80 tubes (100 g) per minute. Nordanmatic is a cartoning machine with a maximum output rate of 150 cartons of toothpaste tubes (100 g) per minute. The morning shift (8 hours) recorded the following production data:

Good output = 39 600 tubes
Routine stoppages = 56 minutes
Other recorded stoppages = 132 minutes
Calculate the OEE of the packaging line.

(see Appendix 3 to compare answers)

8. The causes of the stoppage time of a toothpaste packaging line were recorded as follows:

A: Meal break:	30 minutes
B: Loading of tubes:	10 minutes
C: Loading of cartons:	16 minutes
D: Shortage of toothpaste:	22 minutes
E: Motor failure (Nordanmatic):	54 minutes
F: Jamming of tubes:	12 minutes
G: Jamming of cartons:	24 minutes
H: Other causes:	20 minutes
I: Total stoppage time:	188 minutes

Draw a Pareto Diagram showing the relative share of the causes of the stoppage time.

(see Appendix 3 to compare answers)

9. The activities in a project, their duration and dependencies are shown in the table below:

Activity	Duration (weeks)	Dependency
A	3	–
B	3	A
C	4	B
D	3	A
E	3	C and D
F	2	D
G	4	A
H	4	F and G
I	3	E
J	5	H
K	3	I and J

a. Draw a complete Activity Network Diagram to determine the project duration and critical path
b. Draw a Gantt Chart showing all activities and their duration.

10. a. What are the advantages and disadvantages of Earned Value Management as a tool for monitoring and controlling the progress of a project?

b. A project with a planned duration of 50 weeks has a budget of £300 000. Using the information given in the table below, calculate the time and cost at completion for Week 18:

Week	6	12	18
Planned Spend	£50 000	£90 000	£140 000
Actual Cost	£55 000	£98 000	£150 000
Earned Value	£40 000	£80 000	£130 000

(see Appendix 3 to compare answers)

c. What conclusions can be drawn about project performance up to Week 18 in the above project?

Part 3

Techniques

Introduction to Chapters 9 and 10

This part of the book deals with the techniques of managing and implementing quality. The way in which the tools are used is the technique of application. Therefore in this part the reader will come across frequent references to relevant tools described in Part 2. For example, the Statistical Process Control (SPC) technique also covers Control Chart and Process Capability Measurement.

We have arranged the techniques in two chapters as:

Chapter 9: Quantitative techniques
Chapter 10: Qualitative techniques

In our assessment the criteria of quantitative techniques are that they require statistical analysis and also used in the advanced applications of Six Sigma. The qualitative techniques rely more upon the synergy of group input and do not always contribute to the building blocks of a Six Sigma programme.

9

Quantitative techniques

Product and service quality requires managerial, technological and statistical concepts throughout all major functions in an organization

— Joseph M. Juran

Introduction

This chapter addresses the advanced techniques of building process knowledge quantitatively, by analysing data, predicting outcome and measuring success. These techniques are based upon the quantitative and statistical analysis of data or results derived from other tools. There are a considerable number of quantitative techniques and we have selected the most frequently used techniques in Six Sigma and Operational Excellence programmes as follows:

Q1. Failure Mode and Effects Analysis (FMEA)
Q2. Statistical Process Control (SPC)
Q3. Quality Function Deployment (QFD)
Q4. Design of Experiments (DOE)
Q5. Define, Measure, Analyse, Improve, Control (DMAIC)
Q6. Design for Six Sigma (DFSS)
Q7. Monte Carlo Technique (MCT)

Selection of techniques

Advanced techniques often generate high expectation of results, but there is also a greater danger of using such a technique in a blinkered manner. For each technique, we have pinpointed its specific application areas as well as its benefits and pitfalls. Another reason of grouping these techniques together is that these are also so-called Six Sigma techniques. While selecting these techniques there are some common factors which should be considered with great care by the potential users of these techniques:

- There must be resources of appropriate skills and motivation in the organisation to gain expertise in these techniques.

- The fundamental rigours in purpose, measurement and use must be adhered to.
- The potential benefits, cost and difficulties in using the technique must be assessed.

Structure of presentation

The following section covers the details of each technique under a common structure of presentation as follows:

- *Background*
- *Definition*
- *Application*
- *Basic steps*
- *Worked-out examples*
- *Benefits*
- *Pitfalls*
- *Final thoughts*

Q1: Failure Mode and Effects Analysis

Background

The technique of Failure Mode and Effects Analysis (FMEA) was developed by the aerospace industry in the 1960s as a method of reliability analysis. An early application was found at Allied Signal Turbochargers. In 1972, the Ford Motor Company used FMEA to analyse engineering design and ever since, Ford have refined FMEA through continuous use including its application to Six Sigma. FMEA has proven to be a useful technique of risk analysis in defence industries.

Definition

FMEA is a systematic and analytical quality planning technique at the product, design, process and service stages assessing what potentially could go wrong and thereby aiding faulty diagnosis. The objective is to classify all possible failures according to their effect measured in terms of severity, occurrence and detection and then find solutions to eliminate or minimise them.

Application

There are five basic areas where FMEA can be applied. These are Concept, Design, Equipment, Process and Service.

Concept: FMEA can be used to analyse a product, system or its components in the conceptual stage of the design.

Design: FMEA is applied to analyse a product before the mass production of the product starts.

Equipment: FMEA can also be used to analyse an equipment before it is procured.

Process: With regard to process, FMEA is applied to analyse the manufacturing, assembly and packaging processes.

Service: FMEA can also be applied to test industry processes for failure prior to their release to market.

Basic steps

FMEA involves a 12-step process.

1. Form a team and flow chart the relevant details of the product, process or service that is selected for analysis.
2. Assign each component of the system as a unique identifier.
3. List all the functions each component of the system performs.
4. Identify potential failure modes for each function listed in Step 3. (A failure mode is a short statement of how a function may fail to be performed.)
5. The next step describes the effects of each failure mode, especially the effects perceived by the user.
6. The causes of each failure mode are then examined and summarised.
7. Current controls to detect a potential failure mode are identified and assessed.
8. Determine the severity of the potential hazard of the failure to personnel or system in a scale of 1 to 10.
9. Estimate the relative likelihood of occurrence of each failure, ranging from highly unlikely (1) to most likely (10).
10. Estimate the ease with which the failure may be detected. A scale of 1 to 10 is used.
11. Determine a risk priority number (RPN) for each failure, which is the product of the numbers estimated in Steps 7, 8 and 9. The potential failure modes in descending order of RPN should be the focus of the improvement action to minimise the risk of failure.
12. The recommendations and corrective actions that have been put in place to eliminate or reduce failures are monitored for continuous improvement.

Worked-out example

A worked-out example of a process FMEA from the Ford Motor Company is shown in Figure 9.1. The procedure was used by the Human Resources (HR) department to analyse their 'Internal Job Posting' process. It is interesting to note that the highest RPN was attributed to 'HR Input.'

Failure Modes and Effects Analysis (FMEA)

Process or product name	Internal job posting	Prepared by:	Page .... of ....
Responsible		FMEA Date (Orig)	(Rev) ....

Process step/Part number	Potential failure mode	Potential failure effects	70	Potential causes	Occurence	Current controls	Detection	RPN	Actions recommended	Resp
Finance sign off	Doesn't happen or happens slowly	Stop process	8	Absence of controller; relations with controller; experience of manager. Rules/procedures/bureaucracy	8	Dept/budget headcount limits	1	64		
Grading ratio	Prevent hiring at appropriate grade	Delay in commencing hiring process	8	Ratios out-of-date	7	Grading office/HR function	1	56		
Recruitment request	Delay in receiving request from manager	Ditto	8		1	Company policies	1	8		
Job description	Delayed	Delay in advertising	8	Lack of knowledge by manager. HR rules/procedures	4	HR function	1	32		
Advertising	Does not reach people	No/too few applicants. Quality of applicants. Diversity of applicants. Re-advertise. Delay to process.	7	Company policies – HR/headcount	10	HR/headcount	1	70		
Application	Non/too few. Inconsistent information. Lost Manager refusal to sign	Inability to shortlist/quality of shortlist. Delay to process	7	Advertising process. Headcount/backfill problems/poor paperwork/understanding of process/restrictions on job	7		5	24 5		
Pre-screening	Screen already limited. Number of applicants inconsistent.	Reduce already low number of applicants	2	HR policies	4	HR policies	1	8		
Shortlist/interview	Coordination. No show	Delay to process	5	Poor communication. Availability	2	HR function	3	30		
Selection	No suitable candidates. Candidate rejects offer	Delay to process. Re-advertise	8	Advertising process. Alternative offer	1	None		8		
Feedback	None or variable	Disengagement from process	5	Not priority for HR/manager	3	None		15		
Re-advertise	Make same mistakes as first advert	Ditto	7	Ditto	10	Ditto	1	70		
HR input	Inflexible, inconsistent, inefficient	Delays process	7	Turnover, conflicting priorities, company policies	7	HR function/Company policies. Not customer focused.	9	44 1		
Releasibility/Backfill	Delayed release	Delay in filling post	8	Lack of workforce/succession planning. Headcount restrictions. Project deadlines/departmental	4	20 day rule	1	32		

Figure 9.1 An example of FMEA (adapted from the Ford Motor Company).

Benefits

The major benefits of FMEA include:

- FMEA can be a powerful change agent for identifying weaknesses and risks in a product, process or service and suggesting methods for improvement.
- It is an effective analytical technique to capture the objective components derived from a group work or brainstorming.
- FMEA facilitates the relative weighting of a potential failure before the action is committed at a conceptual or an early stage of an operation.

Pitfalls

FMEA is not an easy technique to handle and has many pitfalls including:

- Writing up the analysis is viewed as a tiresome task.
- The size and experience of the team often create difficulties and the meeting could become 'bogged down.'
- It can be viewed by technical staff as a catalogue of failures and merely a paperwork exercise to satisfy customer agreements.
- The determination of RPN from these factors is often received with doubt and uncertainty. It is viewed as a subjective assessment, as the criteria are not well understood.

Training needs

The members of an FMEA team require 1-day classroom training followed by participation in at least one FMEA real life exercise as observers. The initial reluctance by the manufacturing engineering function to take a leading role in the preparation of FMEA can be overcome by a 1-day 'hands-on' workshop.

Final thoughts

FMEA is conceptually simple, but detail rich. As the process involves steps of writing up modes, effects and causes of failures and assigning numbered ratings subjectively albeit based on experience – FMEA has more sceptics than enthusiasts. Nonetheless, FMEA is effectively accomplished early in the design phase of a new product, process or even a project. Likewise, we do not recommend the use of FMEA when the design has reached a fixed state and changes will be much harder to effect.

Q2: Statistical Process Control

Background

The origin of Statistical Process Control (SPC) can be traced back to the work of Shewhart at Bell Laboratories in the 1920s. During the same period, the late 1920s, a British statistician named Dudding carried out work on statistical quality control along similar lines to that of Shewhart. During World War II, both American and British industries used SPC for the quality control of war materials.

Later in the 1980s, the Japanese led by Tatachi and stimulated by the teachings of Demurg effectively applied the SPC technique in quality programmes. The 1990s witnessed the resurgence of SPC following its successful application in Six Sigma.

Definition

In simple terms, SPC is the control or management of the process through the use of statistical methods and tools.

SPC is about control, capability and improvement and comprises some basic statistics, (e.g. measures of control tendency and measures of dispersion), some tools for data collection (e.g. control charts) and analysis (e.g. process capability).

The seven 'basic tools' as shown are often included as SPC tools, although strictly speaking all of them are not in the SPC category.

- Histogram
- Pareto Chart
- Cause and Effect Diagram
- Check Sheet
- Scatter Diagram
- Flow Charts
- Control Charts

We have described the seven 'basic tools' in Chapter 4.

Application

SPC has four main areas of application:

- To achieve process stability.
- To provide guidance on how the process may be improved by the reduction of variation.
- To assess the performance of a process.
- To provide information to assist with management decision making.

The application of SPC is potentially extensive, ranging from high volume 'metal cutting' operations to non-manufacturing situations including services and commerce.

Basic steps

The objective of SPC is to record the 'voice of the process', remove or reduce 'special' and 'common' causes of variation to the pursuit of continuous improvement. In this context the process means the whole combination of people, equipment, information, input materials, methods and environment that work together to produce output.

The basic steps of a SPC technique are:

- Collect data to a plan and plot the data on a graph such as a control chart or a line graph.
- Use the collected data to calculate control limits to determine whether the process is in a state of statistical control.
- Identify and rectify the 'special causes' of variation to stabilise the process.
- Assess the 'capability' of the process.
- Reduce, as much as possible, the 'common causes' of variation so that the output from the process is centred around a target value.

This is an iterative process in pursuit of continuous improvement. A great deal of effort is required to stabilise the process to be in statistical control, and a great deal more to reduce the common causes of variation.

Glossary

Special causes of variation occur intermittently and reveal themselves as unusual patterns of variation on a control chart. These causes are assignable and are relatively easy to rectify. They include:

- Incorrect material
- Change in machine setting
- Broken tool, die or component
- Keying in incorrect data
- Power failure

Common causes of variation arise from many sources and they do not reveal themselves as obvious or unique patterns of variation and consequently they are often difficult to identify. Some examples of common causes are:

- Poor workmanship
- Poor workplace layout
- Poor quality of materials
- Bad condition of equipment
- Poor operating procedure

Variable data are the result of using some form of measuring instrument, e.g. scale, pressure gauge or thermometer.

Attribute data are the result of an assessment using go/no-go gauges or conforming/non-conforming criteria. An argument in favour of attribute data is that it is a less time-consuming task than that for variables and hence the samples size can be larger.

Arithmetic Mean (χ) is determined by adding all the values and dividing the total by the number of values:

$$\chi = \frac{\Sigma \chi}{n}$$

where Σ means 'the sum of' and n is 'the number of values' in the sample size.

Median is the middle value in a group of measurements when they are arranged in order, e.g. from the lowest to the highest.

Mode is the most frequently occurring value as part of a group of measurements.

Range (denoted by R) is the difference between the smallest and the largest values within the data being analysed.

$$R = \chi_{max} - \chi_{min}$$

Standard Deviation (denoted by σ) is the measure which conveys by how much on average each value differs from the mean.

$$\sigma = \Sigma \sqrt{\frac{(\chi - \chi)^2}{n - 1}}$$

Confidence Level is the degree of confidence or certainty an event is likely to occur for a sample size. The different values of confidence level are given by the area under the curve of normal distribution. To define these values, two attributes are used as shown in Figure 9.2, e.g. χ and σ.

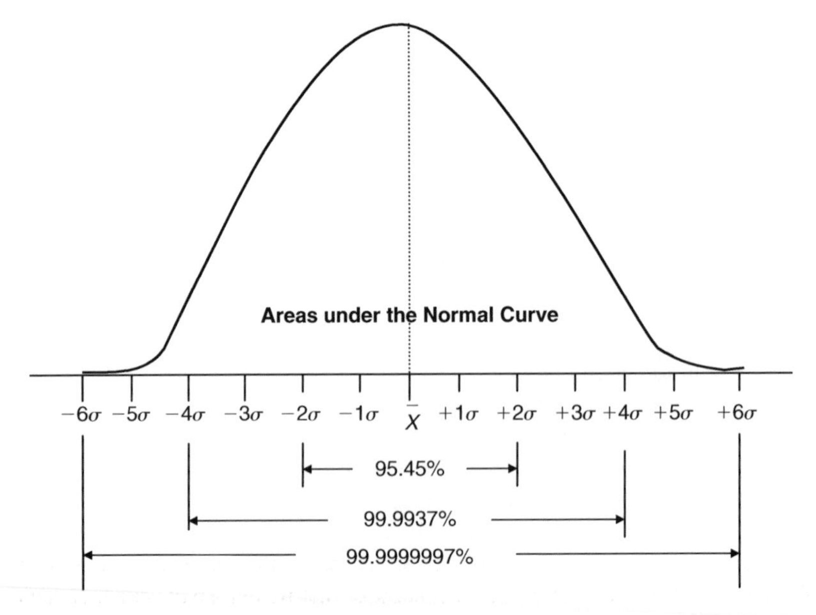

Figure 9.2 Confidence Level.

Central Limit Theorem states that if samples of a known size are drawn at random from a population then sample means will follow a normal distribution, the means of the parent and samples will tend to be the same and the standard deviation of the samples will be approximately that of the population divided by the square root of the sample size.

Population SD = sample SD/SQRT (sample size)

Process Capability refers to the ability of a process to produce a product that meets specification. For example, a highly capable process produces high volumes with few or no defects. The capability of a process is measured by one or a combination of four indices. They are:

DPM = Defects per million
(e.g. for Six Sigma, level of process capability is 3.4 DPM)

σ level = number of standard deviatures between the centre of the process and the nearest specification

$$\tilde{y} = \text{minimum}\left(\frac{\text{USL} - \tilde{y}}{\sigma}, \frac{\tilde{y} - \text{LSL}}{\sigma}\right)$$

where USL = upper specification level
LSL = lower specification level
$\tilde{y}$ = mean
σ = standard deviation

C_{pk} = process capability index
= proposition of natural tolerances (3σ) between the centre of a process and the nearest specification

$$= \text{minimum}\left(\frac{(\text{USL} - \tilde{y}}{3\sigma}, \frac{\tilde{y} - \text{LSL})}{3\sigma}\right)$$

$$= \frac{\sigma \text{ level}}{3}$$

$$C_p = \text{process potential index}$$

$$= \frac{\text{specification width}}{\text{process width}}$$

$$= \frac{\text{USL} - \text{LSL}}{6\sigma}$$

The C_{pk} index takes into account both accuracy and precision and is often defined as a 'process performance capability' index. For a Six Sigma capable process $C_p = 2.0$ and $C_{pk} = 1.5$.

Worked-out examples

Example 1

Determine Mean, Median, Range, Variance and Standard Deviation for the following ordered data set:

3, 4, 4, 4, 5, 6, 6, 7, 8, 30

Mean $\chi = (3 + 4 + 4 + 4 + 5 + 6 + 6 + 7 + 8 + 30)/10 = 7.7$

Median = Average of the two middle values 5 and 6 = 5.5

Range = 30–3 = 27

$$\text{Variance } S^2 = \frac{(3-7.7)^2 + (3-7.7)^2 + (30-7.7)^2}{9} = 63.79$$

$$\text{Standard Deviation } s = \sqrt{63.79} = 7.99$$

Example 2

Calculate C_{pk} and C_p from the following graph (Figure 9.3):

$$\text{Here } \sigma = 2, \tilde{y} = 16, \text{USL} = 20,\ \text{LSL} = 10$$

$$C_{pk} = \text{Min}\left(\frac{\text{USL} - \tilde{y}}{3\sigma}, \frac{\tilde{y} - \text{LSL}}{3\sigma}\right)$$

$$= \text{Min}\left(\frac{20 - 16}{6}, \frac{16 - 10}{6}\right)$$

$$= \text{Min}\left(\frac{2}{3}, 1\right) = \frac{2}{3} = 0.67$$

$$C_p = \frac{\text{USL} - \text{LSL}}{6\sigma} = \frac{20 - 10}{12} = 0.83$$

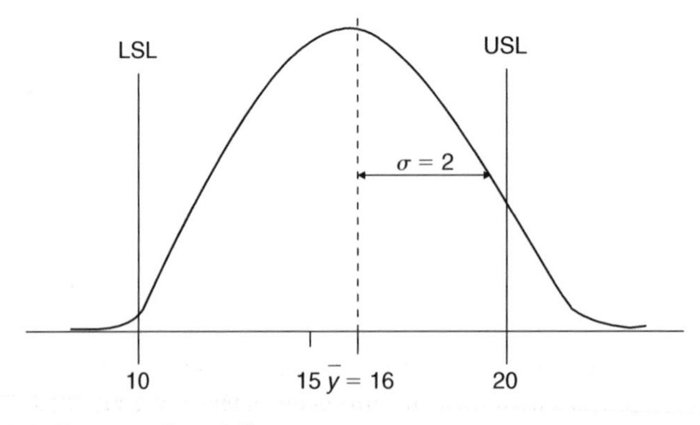

Figure 9.3 Process Capability

Benefits

The major benefits of SPC include:

- The quantitative and statistical foundation of SPC provided the distinction between the TQM approach (outside Japan) of the 1970s and the success of Six Sigma in the 1990s.
- SPC is an effective feedback system that links process outcomes with process outputs leading to continuous improvement.
- The fundamental principle of SPC (i.e. variation from common causes should be left to chance, but special causes of variation should be eliminated) is also a cornerstone of less quantitative approaches of continuous improvement.
- SPC can act as a focal point of a training and company-wide change programme.

Pitfalls

If it is not properly administered with top management support and a facilitator, SPC can be counter-productive. The major pitfalls include:

- Poor understanding of the purpose of SPC within the company.
- Often viewed as 'too much statistics' in shop floor environments.
- Confusion between control and capability, or variation and standard deviation.
- Confusion between seven basic continuous improvement tools and proper SPC tools.

Training needs

The basic understanding of SPC tools requires 1-day classroom training. However, a 3-day workshop is recommended for gaining proficiency in SPC technique. The Six Sigma 'Black Belt' training programme includes an equivalent of 1-week's training in SPC tools and techniques.

Final thoughts

SPC is a powerful technique, but it is basically a measurement process and can make a major contribution to Operational Excellence only when:

- Management is committed
- A structured ongoing training programme is in place
- A mechanism is operational to eliminate 'special causes' and minimise 'common' causes of variation

Q3: Quality Function Deployment

Background

The Quality Function Deployment (QFD) approach was developed by Mizuno and Akao and first applied by Mitsubishi Industries at the Kobe Shipyard. QFD describes a method of translating customer requirements (or the 'voice of the customer') into the functional design of a product or service. The technique has been used particularly by Japanese companies in the 1980s to achieve simultaneously a competitive advantage of quality, cost and delivery. The approach captured the attention of the West after the publication of 'The House of Quality' by Hauser and Clausing (HBR, 1988).

Definition

QFD is a technique that is used for converting the needs of customers and consumers into design requirements and follows the concept that the 'voice of the customer' drives all the company operations.

One specific approach of QFD, called the McCabe approach, proceeds by developing four related matrices in sequence:

- Product Planning Matrix
- Product Design Matrix
- Process Planning Matrix
- Production Planning Matrix

QFD is used to build in quality in the early stage of a new product development and helps to avoid downstream production and product delivery problems. A top level view of QFD is shown in Figure 9.4.

In its simplest form, QFD involves a matrix in which customer requirements are rows and design requirements are columns. When the requirements matrix is expanded by the co-relation of columns the results is called the 'house of quality'.

Application

In congruence with Figure 9.4, QFD is applied to deliver customer needs through four planning phases:

- Product Planning
- Product Design and Development
- Process Planning and Development
- Production Planning and Delivery

During the product planning phase, the deliverables of QFD include customer requirements, competitive opportunities, design requirements and further study requirements.

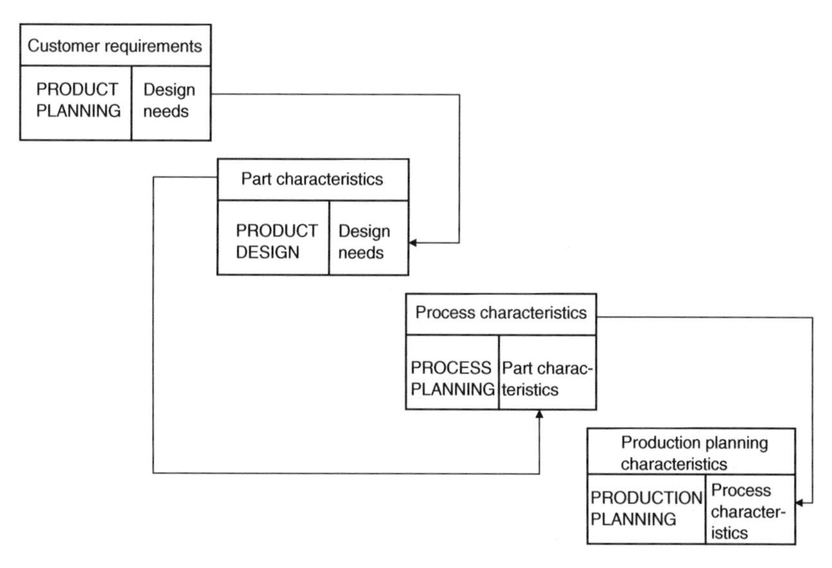

Figure 9.4 QFD approach (© Ron Basu).

The second stage involves product design which translates the design requirements from Stage 1 into component part design characteristics.

The third stage involves the selection of appropriate processes related to specific part characteristics. The deliverable of this stage will be a prioritised list of process characteristics which can be reproduced in production.

The purpose of the production planning stage is to ensure operations plan, training, maintenance and quality plans.

Basic steps

The basic steps of how to develop a 'house of quality' (see Figure 9.5) have been adapted from Mizuno and Akao (1994).

1. *List customer requirements*: The list of customer requirements includes the major elements of customer needs related to a specific product or process. For example, customer requirements for a hotel room include a comfortable bed, a spacious room, a clean bathroom and a television.
2. *List design elements*: These design elements that relate to customer requirements are building materials and specifications. In the context of the hotel room, these are materials used for the mattress, lights in the bathroom, the size of the television and decoration of the room.
3. *Demonstrate relationship between customer requirements and design elements*: A diagram can be used to show these relationships. Furthermore a score on a scale of 1 to 10 can be used; e.g. '9' means 'strongly related'.
4. *Identify the co-relations between design elements*: Positive or negative scores are assigned depending on whether the design elements are positively

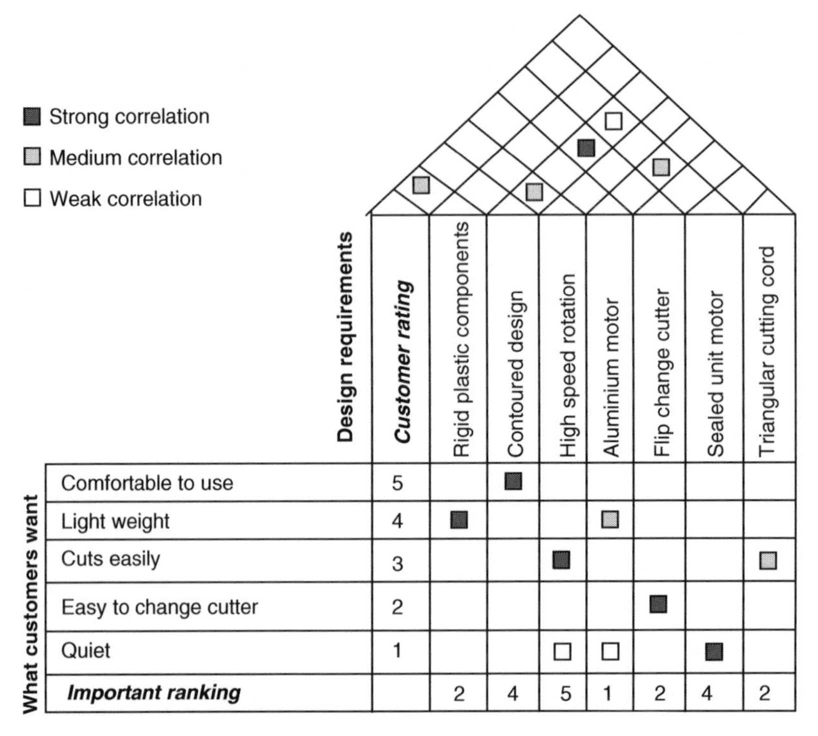

Figure 9.5 The QFD matrix for an electric weed trimmer (© Ron Basu).

or negatively co-related. In the examples concerning hotels, tiles and television are negatively co-related.

5. *Assess the competitiveness of customer requirements*: This is an assessment of how your product (e.g. a hotel room) compares with those of your competitors. A five point scale is used.
6. *Prioritise customer requirements*: 'Customer importance' is a subjective assessment, using a 10 point scale, of how critical a particular customer's requirement is. A focus group of customers usually assigns the rating. Target factors are set on a five point scale (where 1 is for 'no change'). Sales point is established on a scale of 1 or 2 (where 2 is for 'high sales potential'). The absolute weight is then calculated by using the formula:

$$\text{Absolute weight} = \text{Importance} \times \text{Target value} \times \text{Sales point}$$

7. *Prioritise technical requirements*: Technical requirements are prioritised by assessing difficulty level, target value, absolute weight and relative weight. Difficulty level is assigned on a scale of 1 to 10, with 10 being 'most difficult'. Target value is set on a five point scale as before. The value of absolute weight is the product of the relationship between customer and technical requirements and the 'importance of the customer' column. The value of the relative weight is the product of the column of

relationships between customer requirements and technical requirements and the absolute weight of customer requirements.

8. *Final assessment*: The percentage weight factor for each of the absolute and relative weights is then computed. The technical elements with a higher share of relative weight are identified as key design requirements.

Worked-out example

Figure 9.5 shows a worked-out example taken from Waller (2002).

Here the customer requirements for an electric weed trimmer are:

- Comfortable to use
- Light weight
- Cuts easily
- Easy to change cutter
- Quiet

The corresponding technical design elements are:

- Rigid plastic components
- Contoured design
- High speed rotation
- Aluminium motor
- Flip change cutter
- Sealed unit motor
- Triangular cutting cord

The strong correlations between the elements of customer requirements and design requirements are clearly indicated. For example, there is a strong correlation between quiet operation (customer) and sealed unit motor (design). The roof of the 'house' shows high speed rotation (design) has a strong correlation with sealed unit motor (design).

Benefits

The application of the QFD technique has achieved remarkable results, especially in Japan where it has been applied meticulously, to incorporate the 'voice of customers' into all stages of the design, manufacture, delivery and support of product and services. The key benefits of QFD include:

1. QFD seeks to identify the real needs of the customers and translates them into key design requirements.
2. This is a systematic and quantitative procedure which is used to build in quality into the upstream processes and new product development.
3. As the formation of a multi-functional team is a pre-requisite for QFD, this acts as a powerful catalyst to bridge the cultural gaps, especially between marketing and operations.

4. The use of QFD provides a structure for identifying those design characteristics for both products and services that contribute most or least to customer requirements.
5. QFD can be used to link the voice of the customer directly to internal processes. One of the most useful developments in this area is the 'policy deployment' as a measurement based system for continuous quality improvement.

Pitfalls

Although QFD has been a highly successful technique in Japan, there are many difficulties which are experienced during the application of this practice. These pitfalls include:

1. It is often difficult to determine who the customer really is and identify their real needs, especially when in a new market the customer is not certain of his own requirements.
2. The development of matrices which are usually large can be a tedious and exhaustive process.
3. The condition that QFD cannot start until customer needs have been totally defined leads to a delay and loss of motivation.
4. Usually customer data is gathered by marketing, and engineers are eager to finalise the design specifications early. This leads to conflict and often delays the process.
5. Different chart formats are available and there is often a lack of consensus to decide which format would suit the particular project.

Training needs

The basic elements of QFD are not difficult to understand, but its application methodology varies according to the objectives of the project. The team members require attendance in a 1-day workshop. The team leader, unless initially supported by an expert or a consultant, must have hands-on experience of at least one QFD project under his belt.

Final thoughts

When it is effectively used, QFD can bring customer focus to design and shorten the cycle time of the development of a product or services needing fewer changes. However, it is not a magic technique and it requires considerable resources, understanding and attention to detail. It does not require any capital investment but the cost of training and at least six man weeks of time should be taken into account to justify its application. When the application is repeated, the knowledge base is enhanced, cost decreases and benefits increases.

Q4: Design of Experiments

Background

If SPC can be termed as 'listening to the process', we can describe ideas implicit in the Design of Experiments (DOE) as 'interrogating the process'. DOE has become the single most powerful technique in a Six Sigma programme. The origin of DOE dates back to the 1920s when R. Fisher applied complex statistical analysis in agricultural research. The work of Genichi Taguchi on experimental design in the 1970s is regarded as the basis of the current approach for DOE. Taguchi views design from three perspectives: systems design, parameter design and tolerance design.

Definition

DOE is a series of techniques that involves the identification and control of parameters or variables (termed 'factors') that have a potential impact on the output (termed 'response') of a process with the aim of optimising the design or the process. The experiment usually involves the selection of two or more values (termed 'levels') of these variables and then running the process at these levels. Each experimental run is termed as a 'trial'.

As a hypothetical example, consider that you wanted to bake a chocolate cake but you didn't have a recipe to go with it. You need to know how much flour and sugar you require and how many eggs you should use. Furthermore, it would also call for information on the length of time and at what temperature you should bake your cake. A knowledgeable housewife or professional baker would probably resort to trial and error based on experience. However, the use of DOE will provide a systematic approach to finding out the best combination of variables to make your cake.

There are a number of methods of experimentation in DOE, of which the most commonly applied ones are:

- Trial and error method
- Full factorial method
- Fractional factorial method

The trial and error method involves the step-by-step approach of changing one factor at a time, using the experience of the experimenter. This approach is easy to use and understand, but it is inefficient and time consuming.

The full factorial approach considers all combinations of the factors to find the best combination. For examples, three factors with two levels would need 2^3 or 8 trials. Similarly, seven factors with two values will require 2^7 or 128 trials. This method is useful for a lower number of factors.

The fractional factorial method is applied when the number of variables or values is high. Typically for seven factors at two levels, the factorial method would need 32 trials which is a quarter of the full factorial method. This method changes several factors at the same time in a systematic way to ensure the reliability of results.

Application

DOE is an advanced technique which can be applied to both the design of a new product or process or to the redesign of the existing design or process. The technique is most effective for higher levels of variables and values.

The application areas include:

- Product design and process design
- Minimum variation of a system performance
- Reduction of losses in a production line
- Achieving reproducibility of best system performance in manufacture

DOE has become an essential component of an advanced Six Sigma project and is particularly valuable in a Design for Six Sigma (DFSS) project.

Design of Experiment helps to identify the effects of various types of factors as illustrated in Figure 9.6.

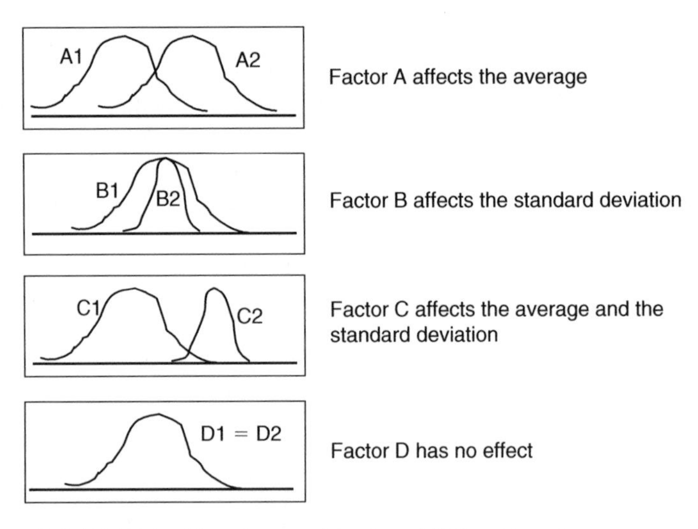

Figure 9.6 DOE identifies types of factors (© Ron Basu).

Basic steps

The steps of a DOE would vary in detail depending on the methods of experimentation. The following basic steps have been simplified to describe the process of orthogonal arrays of the fractional factorial design.

1. *Define the project*: The key information in the project definition report should include:
   - A background statement
   - The purpose, scope and objectives
   - Available statistical data
2. *Develop an IPO diagram*: An IPO (input, process, output) diagram is necessary to establish the critical factors and responses required in the experiment. This step should be undertaken by people who are knowledgeable about the process under investigation.
3. *Select the factors to be optimised*: It is useful at this stage to identify the factors to be optimised during the experiment.
4. *Design the orthogonal array*: The choice of the orthogonal array depends on member demands including the costs of the experiment, the number of factors and the number of levels to be studied. If there are more interactions to be studied, a larger design is required which is usually carried out by computer simulations.
5. *Choose the levels of control factors and the sample size*: The choice of levels is governed by the ease of measurement and the degree of difference between each level. The selection of the sample size and levels are also influenced by the economic consideration of the experiment. One way of reducing the sample size is to identify the highest and lowest level of the 'noise' and to control this.
6. *Carry out the experiment*: It is central that the experiment is carried out by a multi-disciplined team with the rigour of project management principles. The accountability of key tasks such as data collection, leadership and interpretation of results should be clearly defined.
7. *Analyse the data and confirm the results*: The effective relative importance of each factor is determined by the use of variance. For more advanced experiments, Taguchi recommended 'signal-to-noise' ratios for situations requiring different input for a different output. As the orthogonal array is only a subset of the full factorial array it is necessary to carry out a confirmation run.
8. *Close the project*: After the confirmation of the results, the project definition (Step 1) is reviewed to check if the objectives have been met. The methodology, results and conclusions are documented to signal the closeout of the project.

Worked-out example

This example is adapted from Dale (2000) and it concerns the part of the process used in the pharmaceutical industry in the manufacture of tablets.

The objective is to produce uniform tablets in an optimum condition of manufacture. Figure 9.7 shows an IPO diagram of the process.

It was decided to run the first set of experiments to analyse the effect of seven input factors on the particle size on the response at two levels. Table 9.1 shows the experimental layout.

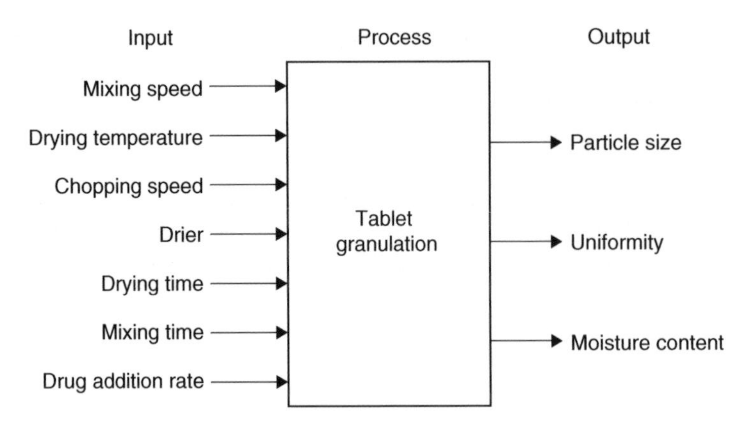

Figure 9.7 IPO diagram.

Table 9.1 Experimental layout

	Control factors	Level 1	Level 2
A	Mixing speed	High	Low
B	Drying temperature	High	Low
C	Chopping speed	Long	Short
D	Drier	Type A	Type B
E	Drying time	Long	Short
F	Mixing time	Long	Short
G	Drug addition rate	Fast	Slow

The results of experimental layout runs with a fractional factorial array (L_8) is shown in Table 9.2. The value of the particle size of each experiment is an average from an acceptable sample size.

The average of the experimental runs for the particle size is 3.96. The values of all responses for each level are calculated as shown in Table 9.3.

Comparisons have been made involving the relative difference between Level 1 and Level 2 of each factor as shown in Table 9.4. The comparative significance of each factor in affecting the response (i.e. Particle Size) is determined by the difference between levels.

From the above analysis it is evident that mixing speed (A), drug addition rate (G), the chopping speed (C) and mixing time (F) have the greatest effect in that order, while other factors have little significance on the response.

In this example, particle size is required to be as small as possible. The values of key factors (A, G, C and F) below the average (3.96) can now be used as the prediction of the result for a combination of factors that reflect their best effect on the output. In this case the optimal factor and levels are A_2, C_1, F_2 and G_1. The other factors can be set at the level where least cost is incurred. The prediction of the optimum combination is shown in Table 9.5.

Table 9.2 Results of experiments

Run	A	B	C	D	E	F	G	Particle size
1	1	1	1	1	1	1	1	3.8
2	1	1	1	2	2	2	2	4.5
3	1	2	2	1	1	2	2	5.3
4	1	2	2	2	2	1	1	4.9
5	2	1	2	1	2	1	2	4.4
6	2	1	2	2	1	2	1	2.9
7	2	2	1	1	2	2	1	2.3
8	2	2	1	2	1	1	2	3.6

Table 9.3 Response values

A_1	=1/4 (3.8 + 4.5 + 5.3 + 4.9)	=18 5/4	=4.625
A_2	=1/4 (4.4 + 2.9 + 2.3 + 3.6)	=13 9/4	=3.300
B_1	=1/4 (3.8 + 4.5 + 4.4 + 2.9)	=15 6/4	=3.900
B_2	=1/4 (5.3 + 4.9 + 2.3 + 3.6)	=16 1/4	=4.025
C_1	=1/4 (3.8 + 4.5 + 2.3 + 3.6)	=14 2/4	=3.550
C_2	=1/4 (5.3 + 4.9 + 4.4 + 2.9)	=17 5/4	=4.375
D_1	=1/4 (3.8 + 5.3 + 4.4 + 2.3)	=15 8/4	=3.950
D_2	=1/4 (4.5 + 4.9 + 2.9 + 3.6)	=15 9/4	=3.975
E_1	=1/4 (3.8 + 5.3 + 2.9 + 3.6)	=15 6/4	=3.900
E_2	=1/4 (4.5 + 4.9 + 4.4 + 2.3)	=16 1/4	=4.025
F_1	=1/4 (3.8 + 4.9 + 4.4 + 3.6)	=16 7/4	=4.175
F_2	=1/4 (4.5 + 5.3 + 2.9 + 2.3)	=15 0/4	=3.750
G_1	=1/4 (3.8 + 4.9 + 2.9 + 2.3)	=13 9/4	=3.475
G_2	=1/4 (4.5 + 5.3 + 4.4 + 3.6)	=17 8/4	=4.450

Table 9.4 Analysis of factors

	A	B	C	D	E	F	G
Level 1	4.625	3.900	3.550	3.950	3.900	4.175	3.475
Level 2	3.300	4.025	4.375	3.975	4.025	3.750	4.450
Difference	1.325	0.125	0.825	0.025	0.125	0.425	0.975
Ranking	1	5	3	7	5	4	2

Table 9.5 Results of optimum combination

	Control factors	Level
A_2	Mixing speed	Low
C_1	Chopping speed	High
F_2	Mixing time	Short
G_1	Drug addition rate	Fast
B_2	Drying temperature	Low
E_2	Drying time	Short
D_1/D_2*	Drier	Type A/Type B

*Depending on lower operating cost

As the orthogonal array is a subset of the full factorial array (e.g. 8 out of 128) a confirmation run is carried out to validate the predicted results.

Benefits

DOE has been recognised as the most important tool for gaining process knowledge. The traditional approach of Cause and Effect Analysis has been to hold all variables constant except one, but in practice it is not possible to hold all other variables constant. DOE has changed that by varying two or more variables simultaneously and obtaining multiple results under the same experimental conditions. In addition to this power of analysing practical conditions, the benefits of DOE include:

1. A properly designed experiment enables you to use the same measurement to estimate several different effects.
2. It provides measurements of process performance and predictability.
3. It pinpoints the opportunities for improvement and indicates where to devote optimum results.
4. Experimental error is quantified and a conformity run validates the conclusions.
5. It enables reproducibility of best systems performance in manufacture by minimising the variation.

Pitfalls

The main disadvantage of DOE is its apparent complexity which distracts potential users. In spite of its extensive deployment in Six Sigma projects, it is still viewed as an academic technique or guarded by 'Black Belt' experts. There are some practical pitfalls involved in DOE beyond this knowledge gap including:

1. The experiments are often run as an isolated intellectual exercise without a multi-discipline involvement or the rigour of project management.
2. For a practical problem with a large number of factors and corresponding responses, it is almost impossible to analyse results without an appropriate software (such as Minitab™).
3. Although considerable process knowledge is gained by the experiment, savings are often difficult to quantify and implement.

Training requirements

The training courses for DOE are usually covered over 1 week. During the course, the participants learn about statistical techniques for systematically manipulating many variables to discover the major factors affecting a selected result variable. The course is taught in a computer lab with each participant assigned to a desktop computer. The participants use statistical proprietary software such as Minitab™, KISS, JMP or a customised excel spreadsheet.

A DOE project is expected to be led by a trained Black Belt or Master Black Belt.

Final thoughts

DOE is not a technique for an amateur enthusiast. If a DOE project is not properly run by a well trained team, there is a danger of finding a solution which is 'exactly wrong rather than approximately right'. However, taking into account this cautionary note, DOE is a powerful technique to build quality into the upstream process by optimising critical characteristics. We recommend the use of DOE as an advanced technique for Operational Excellence invariably supported by a statistical software such as Minitab™.

Q5: Define, Measure, Analyse, Improve, Control

Background

During the early part of the 1980s, Total Quality Management (TQM), especially in the United States and Europe, concentrated on the application of quality management principles and SPC tools within all aspects of the organisation. The focus was upon their integration with key business processes. There were considerable successes in communication and culture but despite this something was missing in the rigour of project management principles. DMAIC (Define, Measure, Analyse, Improve, Control) was introduced by Motorola as the life cycle discipline for Six Sigma projects in the late 1980s. Since then, DMAIC has become the essential component of all Six Sigma initiatives and training programmes. In order to emphasise the sustainability of gains during the Control phase, some companies, (e.g. Dow Chemical) have added the process of Leverage (L) after Control and thus extended DMAIC to become DMAICL.

While Demming's PDCA (Plan, Do, Check, Act) cycle has been extensively used in the development and deployment of quality policies, DMAIC has added the rigour of project life cycle to the implementation and closeout of Six Sigma projects. Figure 9.8 shows the relationship between DMAIC with PDCA and a typical project life cycle.

Definition

DMAIC refers to a data driven life cycle approach to Six Sigma projects for improving processes and is an essential part of a company's Six Sigma programme. DMAIC is an acronym for five interconnected phases: Define, Measure, Analyse, Improve and Control. The simplified definitions of each phase are:

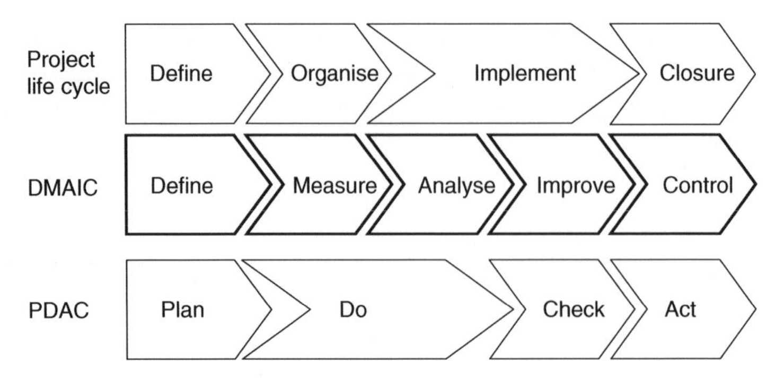

Figure 9.8 DMAIC life cycle (© Ron Basu).

Define by identifying, prioritising and selecting the right project.
Measure key process characteristics, the scope of parameters and their performances.
Analyse by identifying key causes and process determinants.
Improve by changing the process and optimising performance.
Control by sustaining the gain.

Application

The tools of Six Sigma and Operational Excellence, as described in this book, are applied most often within the framework of DMAIC. As such, DMAIC is an integral part of a Six Sigma initiative.

DMAIC is also used to create a 'Gated Process' for project control. For example, the criteria for a particular phase are defined and the project is reviewed and if the criteria are met then the next phase starts (see Figure 9.9).

As a summary of the application of the DMAIC technique, if you cannot define your process you cannot measure it. That means if you cannot express the data you are not able to utilise DMAIC in your development actions. Therefore you cannot improve and sustain the quality.

Basic steps

The key steps of the DMAIC process are well defined. Each phase is required to ensure the best possible results. The process steps are:

1. *Define*: In this initial phase, it is important to define the purpose, scope, objectives and expectations, resources and time line of the project. The customers, their critical-to-quality (CTQ) issues and the core business process involved are clearly identified at this stage.

The tools used for 'Define' include Benchmarking, Project Charter, Process Mapping and SIPOC.

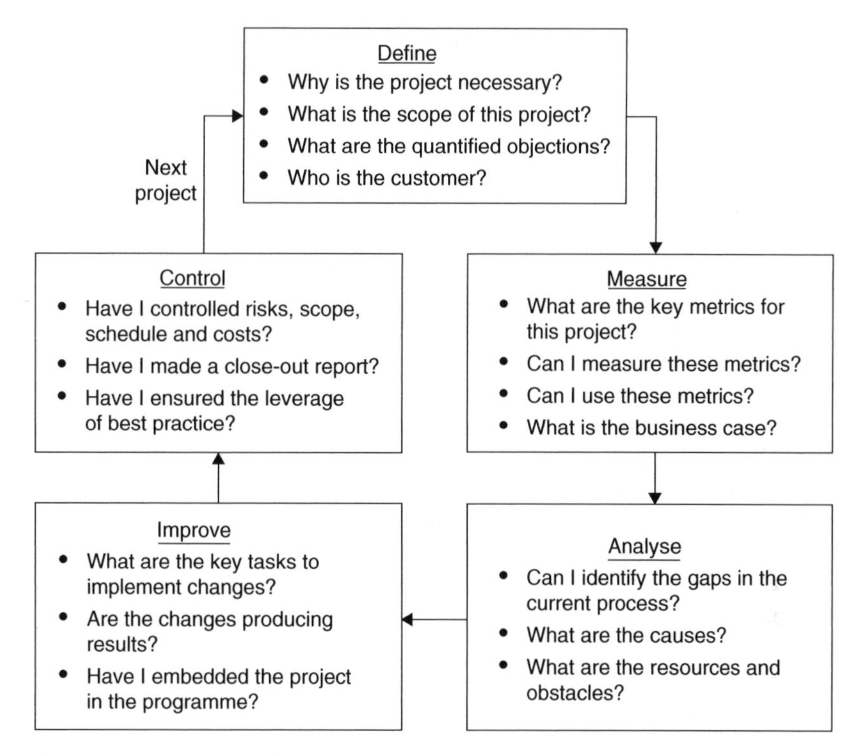

Figure 9.9 The DMAIC cycle as a 'Gated Process' (adapted from www.pyzdek.com).

The deliverables of this phase are:

- Project definition report (Project Charter)
- Customers and CTQs defined
- Business process mapped
- Fully trained team is formed

2. *Measure*: At the start of a journey, the most important thing to know is where you are going. 'Define' provides this road map. However, even if you have a reliable road map you will get lost unless you know where you are now. 'Measure' pinpoints your current location by defining the baseline. One or more CTQs are measured to elicit sufficient data from the process under investigation. At a minimum consider the mean performance and some estimate of variation.

The tools at this stage are Process Flow Charts, selected SPC tools, Pareto Analysis and Run Charts.

The main deliverables of this phase include:

- A proven set of metrics to measure the project progress and objectives.
- Process capability and Sigma baseline.

3. *Analyse*: Once the baseline performance is documented, the objective of the Analyse stage is to validate the root causes of problems. You should be able to compare the various options to determine the most attractive alternative. Striking the appropriate balance between details of the analysis is a key success factor.

Any number of tools and tests can be used. The useful tools for analysis are the Cause and Effect Diagram, SPC tools and the 'Five Whys'. For an advanced application, FMEA and DOE techniques are used extensively.

The important deliverables of this phase are:

- Special and common causes of variance
- Prioritise opportunities to improve

4. *Improve*: During the 'Improve' step, ideas and solutions are put to work. There must be checks to ensure that the desired results are achieved and some trials and experiments may be required in order to find the best solution.

The tools and techniques at this stage include Mistake Proofing, DOE and simulation.

The main deliverables of this phase are:

- The optimum process for improvement
- The key tasks required (work breakdown) to implement the improvement.

5. *Control*: At the final 'Control' phase the stability and reliability of the process is assured and the final capability is determined. The documentation for leveraging the best practice is also part of the Control phase.

The tools and techniques used at this step include SPC, FMEA and Benefits Tracking.

The main deliverables are:

- Cost benefit analysis
- Closeout report

Worked-out example

Although DMAIC is a data driven and quantitative technique, it is also a methodology that may not produce unique output from a given set of input data. We have therefore chosen an actual case (from Dow Chemical) to illustrate a worked-out example.

Define

The Dow Chemical Company began using Six Sigma in 1998 to improve the operations of its subsidiary the FilmTec Corp. of Minneapolis, a manufacturer of membranes that was having difficulty in meeting customer demand.

Membrane quality is determined by two criteria: (a) flux or the amount of water the membrane lets through during a given period and (b) how much impurity is removed from the water.

Measure

Membrane elements were tested prior to shipping to ensure the prescribed quality standards. The specifications were assured but the speed with which his customers were serviced suffered. The shortfall of customer service was costing FilmTec approximately $500 000 a year.

Analyse

Six Sigma tools and methodologies were used at FilmTec to reduce product and process variation. Statistical analysis was used to identify variables that affected membrane flux most significantly.

One of the variables identified for improvement was the concentration of a chemical component used in the manufacturing process. The problem stemmed from the inconsistencies in concentration caused by the interruptions in feeding the chemical in the manufacturing process. The feeding was done from a movable container that was replaced every day. When the empty container was replaced by a full one, the chemical did not always reach the process area. Empty containers were often not noticed.

Improve

To reduce the variation, an inexpensive reservoir was added to feel the chemical while containers were exchanged. Additionally a level transmitter with an alarm was installed to alert operators to containers that were nearly empty.

The improvements have been significant. For one of FilmTec's water membranes, the standard deviation of the product out of specification was reduced from 14.5% to only 2.2%. This resulted in several benefits to FilmTec and its customers including the bonus that membranes were made available to customers faster than before.

Control

To sustain the gains from the project, the Six Sigma team made additional changes. Prior to DMAIC, numbers used to track trends were displayed in tabular form. However, the tables were difficult to read and had little impact on monitoring. Now measurements are displayed on Excel charts that illustrate graphically and with visual immediacy the trends in flux on the finished membrane. The results and learnings from this project were incorporated into a report and posted on the knowledge management website of Dow Chemical.

Benefits

The success of DMAIC is so embedded in the success of Six Sigma initiatives that DMAIC is often regarded as being synonymous with Six Sigma. The training programmes (e.g. Black Belt and Green Belt) are also centred around DMAIC. The grouping of the tools in this book is also roughly structured along DMAIC. The main benefits of DMAIC include:

1. It provides a systematic approach which is common to everyone involved in the Six Sigma programme.
2. DMAIC represents the life cycle of each Six Sigma project – thus imparting a disciple and rigour of project management to achieve its objectives.
3. The results-orientated approach of DMAIC ensures the tracking of benefits and savings.
4. Unlike other advanced quantitative techniques of Six Sigma (viz. DOE, QFD), DMAIC is easy to follow for all members of the project team.

Pitfalls

DMAIC is not immune to pitfalls, however. These include:

1. The critics of Six Sigma often argue that there is nothing new in DMAIC. Similar results, they point out, could be achieved by Demming's PDCA cycle or the basic steps of classical Industrial Engineering.

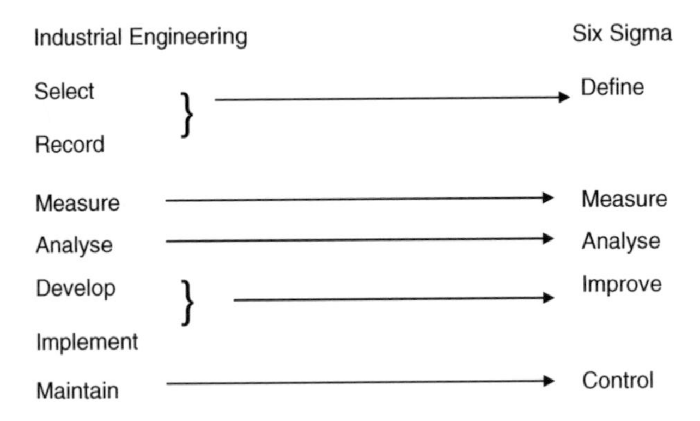

2. It is conceivable that some less complex projects (or the so-called 'low hanging fruits') may be subjected to the rigour of DMAIC more than it is necessary and thus delay the realisation of results.

Training requirements

The training process of DMAIC can be carried out at three levels:

1. Awareness of DMAIC principles in a 1-day workshop.
2. Detailed training of DMAIC (for Green Belts) in 1 week.

3. Training and application of DMAIC as part of Black Belt training, over 5 or 6 weeks. This would include hands-on training of DOE, FMEA and QFD.

Final thoughts

DMAIC is an integral part of Six Sigma. It is systematic and fact based and provides a rigorous framework of results-orientated project management. The methodology may appear to be linear and explicitly defined, but it should be noted that the best results from DMAIC are achieved when the process is flexible, thus eliminating unproductive steps. An iterative approach may be necessary as well, especially when the team members are new to the tools and techniques.

Q6: Design for Six Sigma

Background

The concept of 'Design for Six Sigma' (DFSS) was coined by Motorola who first applied it to the design and production of its pagers in the late 1980s. General Electric initially applied DFSS as a sequel to their Six Sigma programme to move the improvement process one step further, but the process actually turned out to constitute one step back – eliminating the flaws of the product and the process during the Design stage. Many other companies including Dow Chemical, Caterpillar and Seagate Technology have applied and developed the DFSS process further. In Six Sigma, DMAIC has been accepted as a standard methodology. However, in DFSS there is little consistency among terms that define the process. The acronyms range from DMADV (Define, Measure, Analyse, Design, Verify) to DMEDI (Define, Measure, Explore, Develop, Implement) to IDDOV (Identify, Define, Develop, Optimise and Verify), as well as some others.

One fundamental characteristic of DFSS is the verification which differentiates it from Six Sigma. The proponents of DFSS are promoting DFSS as a holistic approach of re-engineering rather than a technique to complement Six Sigma.

DFSS is also known as the application of Six Sigma techniques to the development process. Six Sigma is primarily a process improvement philosophy and methodology while DFSS is centred on designing new products and services. In practice, the difference between a formal DFSS and a Six Sigma programme can be indistinct, as a specific 'Black Belt' project may require DFSS to improve the capability (rather than performance alone) of an existing design.

Definition

DFSS is the system or process of designing and creating a component. This is done with the aim of meeting or exceeding all the needs of customers and the CTQ output requirements when the product is first released. The goal of

DFSS is 'right first time' so that there can be no manufacturing or service issues with the design after the initial release.

Most Six Sigma tools are applicable to DFSS, however the purpose and particularly the sequence of using these tools can be different from what might be expected within an ongoing operation. For example, DOEs are often used in Six Sigma to solve problems in manufacturing, when in DFSS structured fractional factorial DOEs could be used as a control measure to achieve a high quality product design. A large DFSS project requires the use of advanced techniques, such as QFD, FMEA and DOEs.

Application

The primary application of DFSS as a technique is in the design and development stage of a product, process or service.

There are examples of the successful application of DFSS (Choudhary, 2003) in three environments:

1. Business transactions and services
2. Manufacturing processes and products
3. Engineering products

In the service industries, DFSS methodology bypasses the Measure and Analyse phases of DMAIC by creating a process that prevents the variations from emerging. This prevention methodology moves the five Sigma performances to yield Six Sigma results.

In manufacturing processes and products, DFSS encompasses the methodology of QFD and gets the designers and contractors working in concert to optimise the capability of the manufacturing process to attain a consistent quality for the product.

With regard to engineering products, in addition to applying the principles of concurrent engineering which entails the development of both products and processes, DFSS uses DOE for design optimisation.

Basic steps

There are a number of process steps for putting DFSS into practice of which the most frequently reported methodologies are DMADV (Define, Measure, Analyse, Design and Verify) and IDOV (Identify, Design, Optimise and Validate). DMADV is often described as the next stage of DMAIC and thus may lead to a generic approach. In order to emphasise the distinctive characteristics of DFSS we have adapted IDOV to show the basic steps of the process.

1. *Identify Phase (Define and Measure)*: The Identify Phase begins with the development of a team and a formal link of the design to the 'voice of the customer'. The fundamental rule is; do not start a DFSS project with out

the customer, management commitment and marketing involvement. The team members should have extensive experience in Six Sigma tools and techniques.

The key tasks of this phase, also known as the 'Measure Phase', are:

- Identify customer and product requirements
- Identify CTQ variables and technical specifications
- Establish a business case
- Establish the roles and responsibilities of team members
- Plan the project and set milestones

The most common and effective technique at this stage is QFD.

2. *Design Phase (Analyse)*: The Design Phase consists of developing alternative concepts, evaluating each option and selecting the best-fit concept. This is also known as the Analyse Phase where design parameters (CTQs) are deployed. The key tasks of this phase include:
   - Formulate conceptual design
   - Identify potential risks by using FMEA
   - Use DOE to determine CTQs and their influence on the technical requirements
   - Develop a bill of materials, formulation and procurement plans
   - Outline a manufacturing plan
   - DOE and FMEA are the most appropriate techniques for the Design Phase
3. *Optimise Phase (Design)*: The Optimise Phase uses the process capability information and simulation to develop and optimise detailed design elements. This is also called the Improve Phase and performance is predicted at this phase. The key tasks include:
   - Assess process capabilities of each of the design parameters to meet CTQ limits
   - Use simulation to predict and design for robust performance
   - Optimise tolerance and cost
   - Implement design, commission and start up

The key tools used at this stage are SPC tools and simulation

4. *Validate Phase (Verify)*: The validate phase comprises the testing and validation of the design. During this phase, also known as the Control Phase, feedback of requirements and possible changes are shared with the design, manufacturing and procurement teams. The key tasks at this stage include:
   - Test prototype and validate
   - Assess failure modes, performance and risks

The FMEA technique is extensively used at this stage.

Worked-out example

The following flow diagram (Figure 9.10) is used as a worked-out example for DFSS. The diagram illustrates the main steps of developing a new consumer product by using DFSS.

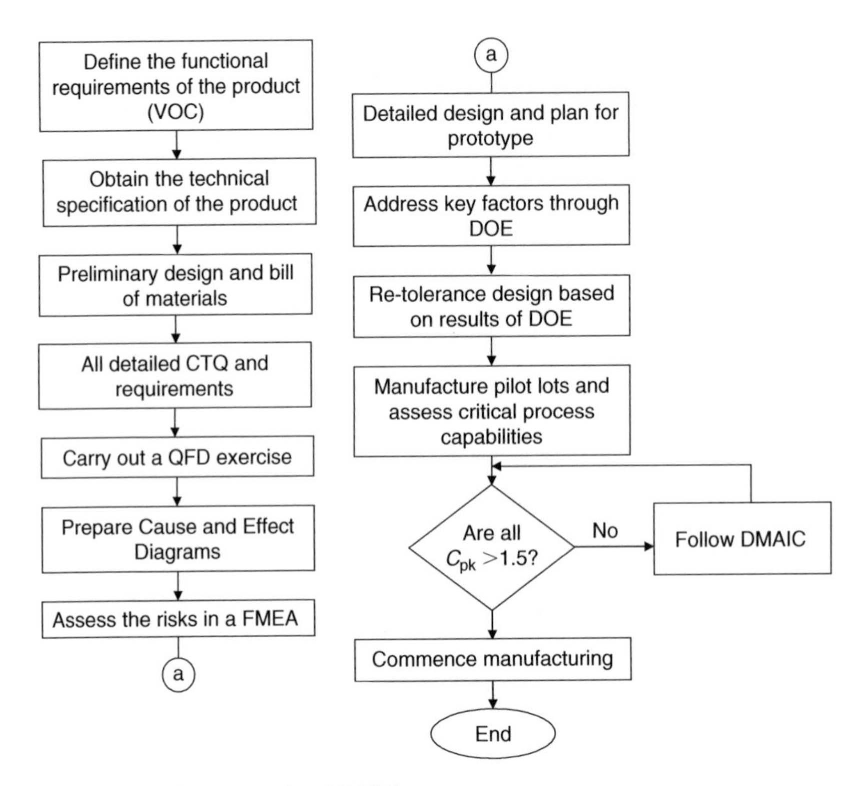

Figure 9.10 An example of DFSS.

Benefits

The proponents of DFSS and successful DFSS projects have established that DFSS helps fulfil the 'voice of the business' by fulfilling the 'voice of the customer'. The key benefits can be summarised as:

1. DFSS satisfied the voice of the business by:
   a. Increased sales volume by opening new markets with new designs and decreasing the cost of an existing design
   b. Decreasing development cost and capital investment
2. DFSS satisfies the voice of customers by:
   a. Improved design of existing products
   b. Generating value through new products
   c. Reducing the time to deliver new products

3. DFSS helps to improve organisation effectiveness by:
   a. Generating a discipline for product development excellence
   b. A multi-functional synergy through active leadership to a common goal.

Pitfalls

A major drawback of DFSS is the emphasis by its proponents on a 'pure play' DFSS project. Thus it is often misunderstood by users and sponsors. Its links with other relevant techniques and methodologies, such as Six Sigma, DMAIC, QFD and concurrent engineering, are either confused or over-emphasised.

There are some practical pitfalls beyond the conceptual or philosophical issues such as:

1. The costs of the projects are more visible than the benefits for a long time along the project life cycle.
2. A DFSS project usually runs for a long period and the project closeout is not always well defined.
3. Based upon cost benefit alone, the justification of a DFSS is difficult, especially for a new product with an unpredictable forecast.
4. The success of DFSS depends heavily on the knowledge of the project team regarding advanced techniques such as QFD, DOE and FMEA.

Training requirements

A DFSS programme requires that it is led by a trained Black Belt or Master Black Belt. The team members should also have a good training and hands-on experience in key Six Sigma tools and techniques, particularly in QFD, FMEA and DOE. If the trained team leader or members are not readily available, we recommend that a well structured 1-week training programme is put into place before the start of a DFSS project.

Final thoughts

DFSS is a longer-term, resource-hungry process and it is expensive. Therefore, it should be deployed with care and on just a few vital projects, and specifically targeted towards the development of new products. Do not start a DFSS project without the customer, sales involvement, top management commitment and a team, preferably one with Six Sigma training. DFSS is a powerful technique and its power should not be abused.

The proponents of DFSS believe that within the new few years as experience grows, DFSS will be used in Design Houses with the same familiarity as ISO standards. DFSS is primarily for new process or product development. Figure 9.11 illustrates the relationship between DFSS (DMADV) and DMAIC.

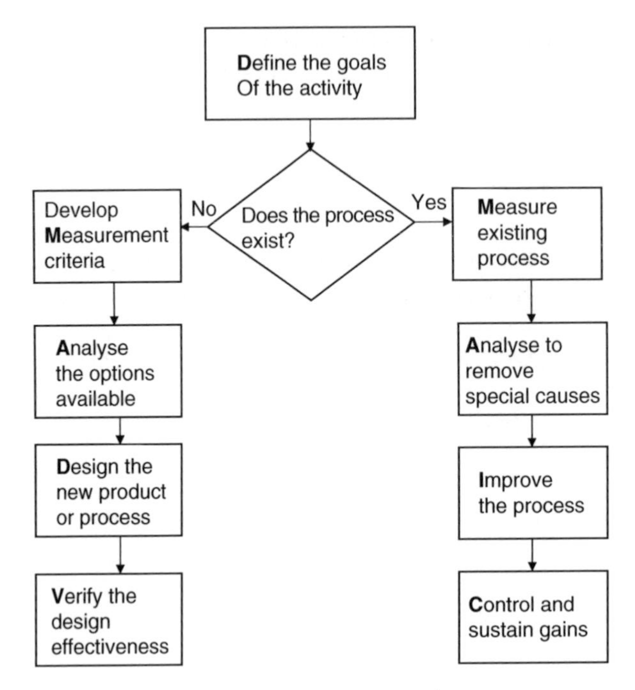

Figure 9.11 DFSS versus DMAIC (© Ron Basu).

Q7: Monte Carlo technique

Background

The origin of the Monte Carlo technique (MCT) goes back to a legendary mathematician observing the unpredictable behaviour of a saturated drunk. Each of the drunk's steps was supposed to have an equal probability of going in any direction. The mathematician wanted to estimate the average number of steps the drunk had to take to cover a specified distance. This was named the problem of 'random walk' and thus the application of random sampling was born, as the story goes. The method was found to have many practical applications, and subsequently it was named the Monte Carlo technique. The method requires a random number producing device, such as a roulette wheel, hence the name 'Monte Carlo'.

Definition

Churchman et al. (1968) define the MCT as 'simulating an experiment to determine some probabilistic theory of population of objects or events by the use of random sampling applied to the components of the objects or events'.

It often happens that an equation is too complex that it does not yield a quick solution by standard numerical methods. However, a stochastic process with distributions and parameters which satisfy the equation may exist. Instead of using the pureplay method it may be more efficient to construct the stochastic model of the problem and compute the solution. Thus an experiment is set up to mirror the features of the problem and a simulation is carried out by supplying random numbers into the system to obtain numbers from it as an answer. This is the basis of the Monte Carlo Simulation process. The method does not mean a 'gamble', as the name may imply, but rather it refers to the manner in which individual numbers are selected from valid distribution functions. As shown in Figure 9.12, the output from a Monte Carlo simulation includes a histogram showing the relative probability of events and a cumulative probability chart ('S' curve).

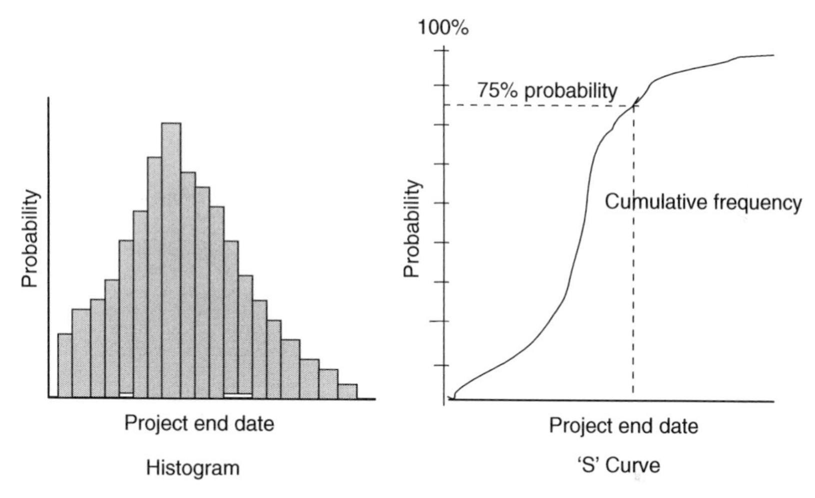

Figure 9.12 Output from Monte Carlo simulation (© Ron Basu).

Application

MCT has been successfully used in scientific applications for over eight decades. The method has been a cornerstone of all NASA projects. It has been used extensively in investment appraisal and risk analyses in both scientific and industrial programmes.

There is an abundance of research publications containing the application of the MCT, particularly in solving queuing problems. It is evident that this technique can be used to solve any queuing problem for which the required data can be collected. Saaty (1959) demonstrates the successful application of the MCT in the optimisation of telephone calls, landing of aircraft, loading and unloading of ships, the scheduling of patients in clinics, customers and taxis at a stand and flow in production, to name but a few. It is important to realise that all these applications were accomplished before the advent of the computer.

Although the MCT is still a popular topic of academic and scientific research, with the availability of proven simulation software (e.g. PRM and

Primavera Monte Carlo), the system is now in wider use in businesses and project management.

The applications of the MCT in project risk management are focused on two stages of the project life cycles, viz. feasibility and implementation. At the feasibility stage the simulation is the perfect tool for evaluating a pro forma of investment opportunities. We know that at the initial stage of a project there is uncertainty regarding input since pro forma is by definition an estimation of performance. We also know that the MCT will calculate answers that accurately reflect the uncertainty of input data. Figure 9.13 illustrates the probability curves, using Monte Carlo simulations, of the net present value (NPV) for two options of a wind energy project in Denmark.

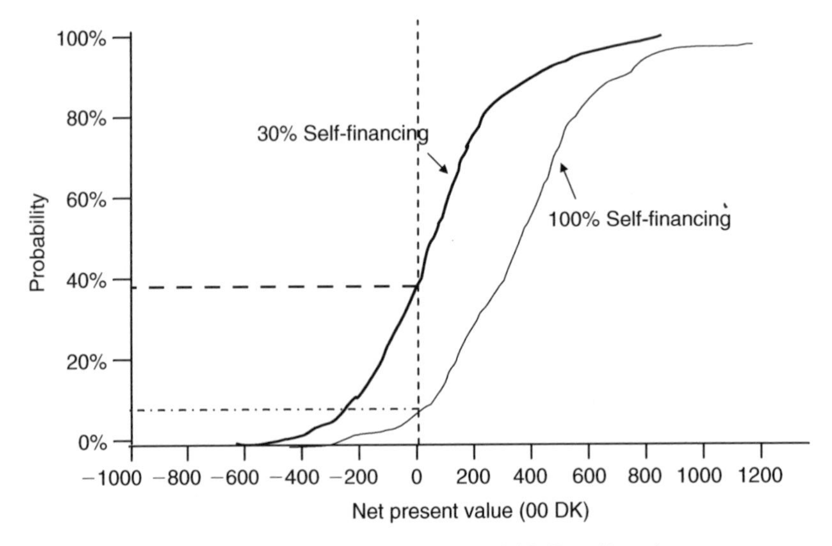

Figure 9.13 Monte Carlo in project appraisal (© Ron Basu).

During the implementation stage of a project, the MCT examines the overall uncertainty of project completion times (see Figure 9.12).

The technique is also very useful in assessing the probability and consequence of critical risk in a risk register during the implementation of a project. For example, having completed the preliminary assessment of probability/impact scores on the risk register for the project cost estimate and schedules, the data for high score risk items are loaded in the Monte Carlo risk analysis software. This provides a quantitative evaluation of cost and schedule for high risk items.

Basic steps

1. Observe some basic rules on the use of this technique including:
   a. The scope of each problem must be identified by studying each process individually to focus on the core issue and the aspects that are ignored.

   b. The strength of the technique lies in its applicability to an equation which demands numerical solution but is not easy to solve by standard numerical methods.
   c. If in a problem there are fluctuations due to time, apply the method to the parts which appear to cluster together in time.
   d. The simulation should be applied several times to a problem with different sample sizes.
   e. The technique follows a normalised distribution function. That means that we multiply the frequencies of the distribution functions by a constant such that the area under the normalised distribution function is equal to 1.
2. In practice, a Monte Carlo simulation is carried out by a computer software. The input data requirements for each variable are the estimates for:
   a. Minimum value
   b. Most likely value
   c. Maximum value

By using random numbers and sampling techniques, the computer system will generate a probability distribution of each parameter. From the cumulative probability charts ('s' curves), the probability level is decided and the corresponding value of the parameter is selected.

3. The iterative process of the MCT is described below. (This can be done manually, although the use of a computer system is more sensible.)
   a. Given the value of X_i lies between a range of X_{min} and X_{max}.
   b. Start the *i*th iteration. Use a Random Number between 0 and 1 and the normalised Distribution Function of X_i to determine the single value of x_i to be used in this iteration.
   c. Place the value of X_i in the appropriate 'bin' of a histogram. This histogram will be come a Frequency Distribution of x_i when all iterations are complete.
   d. Go to Step (b) and start the process over until the Frequency Distribution in Step (c) is complete for the range X_{min} and X_{max}.
   e. Plot the cumulative probability chart or 's' curve for X_i (see Figure 9.12) and decide the probability level and corresponding value of X_i.

Worked-out example

This example is adapted from Churchman et al. (1968) and it concerns the home delivery of packages of goods purchased at a department store.

Consider the packages arriving for delivery from the store:

- Normal distribution
- *Mean arrival rate*: 1000 packages/day
- *Standard deviation*: 100

Similarly, assume the service pattern of delivery by trucks:

- Normal distribution
- *Mean service rate*: 100 packages/day
- *Standard deviation*: 10

Let us also assume, for the purposes of this example:

- Cost per truck: \$25/day
- Cost of overtime: \$8/hour

In practice, there could be a cost of delay, but in this exercise we assume that packages will be delivered on the same day and overtime may be required.

We run a Monte Carlo method of the delivery system 'on paper' with two fleets (10 and 12 trucks each) for five consecutive days. In a computerised simulation we could run the system for any length of time and for as many fleet sizes as desired. We begin by preparing a table as in Table 9.6.

In the above table Columns 1 and 2 are self-explanatory.

Column 3 refers to five successive numbers from the table of Random Normal Numbers (see Appendix 2).

Column 4 shows the converted value of the number of packages arriving by taking into the account the Standard Deviation and Random Normal Number. This is equal to 1000 + 100 × Column 3.

Column 5 is the total requirement = Column 4 + Previous Column 8.

Column 6 is the next set of five consecutive numbers from the Random Normal Number table. This represents the average number of deliveries per truck.

Column 7 is the converted value of total deliveries = Column 1 × (100 + 10 × Column 6).

Column 8 shows the leftover packages if no overtime is deployed. This is equal to Column 5 − Column 7.

Column 9 refers to the number of packages to be delivered at overtime rate. This is equal to Column 4 − Column 7.

Column 10 is the cost of overtime. This is calculated as

$$[\text{Column 9} \frac{/\text{Column 7}}{8 \times \text{Column 1}}] \times \$8.00$$

The total costs per week for each fleet can now be compared. Assuming a five day week, costs per week are:

10 trucks: 10 × 5×25 + 296 = \$1546

12 trucks: 12 × 5×125 + 56 − \$1556

In this case the 10-truck fleet is more economical.

Benefits

1. The MCT is one of the most used simulation methods in both academic research and industrial applications. A major difficulty encountered in

Table 9.6 Manual simulation of a delivery system

1	2	3	4	5	6	7	8	9	10
Trucks in fleet	Day	Random Table Value 1	Packages arrived	Total requirement	Table Value 2	Number of deliveries	Leftover packages	Number delivered at overtime	Cost of overtime ($)
	1	2.455	1246	1246	−0.323	968	278	278	184
	2	−0.531	947	1225	−1.940	806	419	141	112
10	3	−0.634	937	1336	0.697	1070	286	0	
	4	1.279	1128	1414	3.521	1352	62	0	
	5	0.046	1005	1067	0.321	1032	35	0	
Total									296
	1	2.455	1246	1246	−0.323	1161	85	85	56
	2	−0.531	947	1032	−1.940	967	65	0	
12	3	−0.634	937	1002	0.697	1284			
	4	1.279	1128	1128	3.521	1623			
	5	0.046	1005	1005	0.321	1239			
Total									56

operations research is that of dealing with a situation so complex that it is impossible to set up an analytical equation. The MCT aims at simulating the operation systematically where the major factors and their interaction are studied.
2. This technique calculates the answers that accurately represent the input data by a large number of iterations. Experience over the last 80 years has proven that the solutions are also very close to real circumstances.
3. With the support of effective computer software, this technique has become an essential part of risk assessment in various stages of major projects.

Pitfalls

1. The MCT is not free from pitfalls. These include:
2. The methodology and algorithm of the MCT are perceived as very complex for practical managers. The theory of probabilities and the characteristics of random numbers are considered to be the domain of academics and statisticians.
3. Without the use of a computer software, the iterations by a manual process can be laborious and prone to error.
4. Many practitioners tend to shy away from this technique with a notion that 'it is better to be approximately right than exactly wrong'.

Training needs

The application of the MCT requires a good understanding of statistics, and how the technique actually works. It is also necessary for the user to gain hands-on experience in the use of the computer software. We strongly recommend that the modelling and simulation of a practical problem by the MCT should be given to someone who is trained and experienced in simulation techniques.

The project team members should receive an awareness training so that they are happy with the interpretation of the results obtained by the technique.

Final thoughts

The MCT is an excellent aid to risk management, but it should be used with care, ideally by trained specialists and always with the support of an effective computer software.

10

Qualitative techniques

He that will not apply new remedies must expect new evils; for time is the greatest innovator

— Francis Bacon

Introduction

Chapter 9 reported the advanced quantitative techniques of analysing data and improving process performance. In this chapter we shall deal with another category of techniques which we have named 'qualitative techniques'. Although they are data driven, they depend heavily on logical reviews and judgemental assessments. There are numbers, but the improvements are not directly derived from a statistical process or a numeric solution. There are many such qualitative techniques and we have selected the most relevant ones for Six Sigma and Operational Excellence programmes as follows:

R1: Benchmarking
R2: Balanced Scorecard (BSC)
R3: European Foundation of Quality Management (EFQM)
R4: Sales and Operations Planning (S&OP)
R5: Kanban
R6: Activity Based Costing (ABC)
R7: ISO 9000

The selection criteria of these techniques are the same as those for the quantitative techniques, except that for qualitative techniques there is less emphasis on specialist knowledge. However, qualitative techniques require specific training, in particular for S&OP, EFQM and ABC.

We have followed the same structure of presentation as in Chapter 9, having pinpointed specific application areas as well as their benefits and pitfalls.

R1: Benchmarking

Background

The origin of Benchmarking as it is known today, is credited to the Xerox Corporation who applied the improvement method based on comparing performances in the 1980s. The concept of Benchmarking was well publicised by Camp (1989) based on the best practices of the Xerox Corporation.

It is to be noted that although the process was not formalised many organisations, both in the private and public sectors, have been carrying out comparisons of the performance levels of various units for many years. The work of Camp (1989) was followed by numerous publications on benchmarking including Colding (1995) and Karlof and Ostblom (1994). Now various forms of benchmarking and sharing of best practices have become an accepted process for both performance improvement and knowledge management.

Definition

According to Karlof and Ostblom (1994), benchmarking is a continuous and systematic process for comparing your own efficiency in terms of productivity, quality and best practices with those companies and organisations that represent excellence.

Dale (1999) suggests three main types of formal benchmarking:

- Internal benchmarking
- Competitive benchmarking
- Functional benchmarking

Internal benchmarking involves benchmarking between the same group of companies so that best practices are shared across the corporate business.

Competitive benchmarking relates to a comparison with direct competitors to gather data on 'best in class' performance and practices.

Functional benchmarking is a comparison of specific process in different industries to obtain information on 'best in school' performance and practices.

Internal benchmarking is the easiest one of the three to carry out while competitive benchmarking is often the most difficult one to put into place. Organisations are usually keen to share data in functional benchmarking when there is no direct threat of competition.

Application

A benchmarking process seeks to provide knowledge in a number of areas including:

1. What are the potential opportunities for improvement in our products or processes?
2. Who the 'best in class' industry leaders in our competitive market are?
3. How is our performance comparing with those of industry leaders?

The above three objectives could be provided respectively by internal, competitive and functional benchmarking. The application areas usually depend on the type of benchmarking.

Internal benchmarking is widely used by a multi-national business with a number of subsidiaries in different countries or a national business which operates with some kind of branch structure of divisions. In such cases the business contains a number of similar operations that can be compared. GlaxoSmithKline uses internal benchmarking to compare Key Performance Indicators (KPI) and to identify potential cost savings in its business. The Foods Division of Unilever have used internal benchmarking to compare Best Proven Practice for each of its major manufacturing sites.

Establishing a benchmarking partnership with other competitors can be mutually beneficial to both parties for the purpose of positioning the company in the market. External consultants and industry associations often play a key role to set up and conduct benchmarking on competitive issues and opportunities such as e-commerce, purchasing commodity-type materials and customer perceived quality.

The object of functional benchmarking is to identify the best practice wherever it may be found. The purpose is to benchmarking a part of the business which displays a logical similarity even in different industries. For example, a battery manufacturing company may want to benchmark its standard of direct delivery with Dell Computers, an organisation of acknowledged excellence in this area. Similarly, a pharmaceutical company may compare their production line changeover practices with Toyota Motors in Japan.

Basic steps

1. *Identify what to benchmark*: Identifying what to benchmark is influenced by the knowledge of your own business. A SWOT (Strengths, Weaknesses, Opportunities and Threats) analysis may point towards the subject to be benchmarked. The subject could be products, production lines, customer service, working practices and so on.
2. *Plan the benchmarking process*: The preparation and planning should include:
   a. Forming a team with their roles and responsibilities.
   b. Selecting the measures of performance for the selected activity for benchmarking.
   c. Method of data collection.
   d. Defining the scope and time line for the exercise.
3. *Identify benchmarking partners*: The participating units will vary depending on the type of benchmarking. Having identified a number of potential candidates, whether internal or external to the organisation, they must be contacted and briefed regarding the objective of the exercise. It is essential to establish a mutual trust with potential partners otherwise all efforts could be fruitless.
4. *Collect data*: The purpose of the fourth stage is to supply the information needed for the analysis. It is useful to draw up a questionnaire and test it by starting in your own organisation. The collection is often supplemented by interviews with partners.

5. *Analyse data*: A comparative analysis of the validated data is carried out to identify gaps. It is critical that the reasons for the gap with the 'best in class' are determined and understood. The trend of the gap should also be estimated over an appropriate time frame.
6. *Implement and improve plan*: Develop action plan of closing the gap which the analysis stage has identified. The action plan is then implemented often requiring effective project management.
7. *Review and repeat*: More often than not, a benchmarking exercise is a continuous process. This should be conducted on a regular basis by sharing the results with benchmarking partners.

Worked-out example

The following example is taken from Welch and Byrne (2001).

General Electric Inc., with its global business of over US $120 billion per annum has been voted by Fortune as the 'most respected company'. GE is also known as the 'Cathedral of Six Sigma' and the high profile of the programme under the leadership of Jack Welch has been well publicised. GE licensed Six Sigma technology in 1994 from the Six Sigma Academy, rolled out the programme worldwide and achieved $2 billion savings in 1999.

GE Capital is the financial services arm of GE and accounted for approximately 40% of the group's turnover in 2001.

One heartening early success story at GE Capital relates to benchmarking and the sharing of good practice. GE Capital fielded about 300 000 calls a year from mortgage customers who had to used voicemail or callback 24% of the time because employees were unavailable. A Six Sigma team found that one of their 42 branches had a near perfect rate of answered calls.

The team carried out a benchmarking exercise between all branches. They analysed the systems, process flows, equipment, physical layout and stopping of the 'best in class' branch and then cloned it to the other 41 branches. As a result, a customer has a 99.9% change of obtaining a GE person on their first try.

Benefits

The benchmarking techniques had delivered many tangible and intangible benefits. These include:

1. It enables process owners to identify what is to be improved and motivates them by the knowledge of what is achievable.
2. It enhances the communication, trust and partnership spirit between participating units by focusing on the high visibility of key processes and performance indicators.
3. It removes complacency even for high performers by providing a platform for improvement.
4. By focusing on processes rather than individuals, it helps to eliminate defensive and 'finger pointing' practices.

5. It focuses attention on the details of best practice, thereby initiating a process of generating a learning culture.
6. It has a remarkable effect on strategy formulation, strategy implementation and leadership development.

Pitfalls

There are certain difficulties and pitfalls which must be recognised to make the benchmarking technique a success. These include:

1. *Lack of trust in the process*: Unless members at all levels of a participating company believe that the business can benefit, the exercise has little value.
2. *Poor choice of metrics*: The measures for comparison are often poorly chosen without due consideration for local variations. There is also a danger of number games being played or encouraging a 'league table' culture.
3. *Inadequate process understanding*: The exercise tends to focus on the outcome of a process rather than analysing the causes of the outcome.
4. *A paper exercise remote from the source*: Without the full participation of members from each unit the exercise could be perceived as a mere 'paper exercise' by the centre. The centre is then accused of not understanding the variation and complexity of participating units.
5. *Poor communication*: If the participating units are not fully aware of the outcome of the project, data sharing will be guarded and the project is likely to fail.

Training requirements

The team members of participating units require opportunities of team building to develop a mutual trust in sharing information. This is more fundamental than the knowledge of the key principles and steps of benchmarking. One method of achieving this is by encouraging the visits to selected sites involved in the process. This is further supplemented by e-communications and thus forming a virtual team.

Final thoughts

Benchmarking exerts a powerful impact on learning organisations by leveraging best practices. It is a technique for knowledge management and not a cost cutting method or a platform for number games.

R2: The Balanced Scorecard

Background

The concept of the Balanced Scorecard (BSC) was first introduced by Kaplan and Norton (1992) in an article in the Harvard Business Review, *The*

Balanced Scorecard – Measures that Drive Performance. This generated considerable interest for senior business managers and led to the next round of development. The focus was shifted from short-term measurement towards generating growth, learning and value added services to customers. This methodology was then published by Kaplan and Norton in a number of articles in the Harvard Business Review and in a book, *The Balanced Scorecard* (1996).

The senior executives of many companies are now using the BSC as the central organising framework for important decision processes. The rapid evolution of this technique has gradually transformed the performance measurement process into a strategic management system.

Definition

The BSC is a conceptual framework for translating an organisation's strategic objectives into a set of performance indicators distributed among four perspectives: Financial, Customer, Internal business processes and Learning and growth (see Figure 10.1).

The indicators are aimed to measure an organisation's progress towards achieving its vision as well as the long-term drivers of success. Through the BSC, an organisation monitors both its current performance (e.g. internal processes, finance, customer satisfaction) and its effort to improve and sustain performance (e.g. innovation and employee development). It is also balanced in terms of internal efficiency and external effectiveness.

For further details on the BSC see Kaplan and Norton (1996).

Application

The BSC has been applied successfully in several organisations around the world. The scorecard, with some customised changes, provides a management tool for senior executives primarily to focus on strategies and longer-term objectives. The organisations could vary from a large multi-national business to a non-profit making public service unit. The scorecard is sometimes named the 'Executive Dashboard'. The KPI are reported as:

- Current actual
- Target
- Year-to-date average

When the actual performance value is on or above target then the value is shown as green. If the actual is below the target but within a given tolerance then the colour becomes amber. It is depicted in red when the value is below the tolerance limit of the target.

Another area of application is to assess the performance at the tactical operation level. Usually the top level indicators (also known as 'Vital Few') are designed in such a way that they can be cascaded to 'component' measures and the root causes can be analysed.

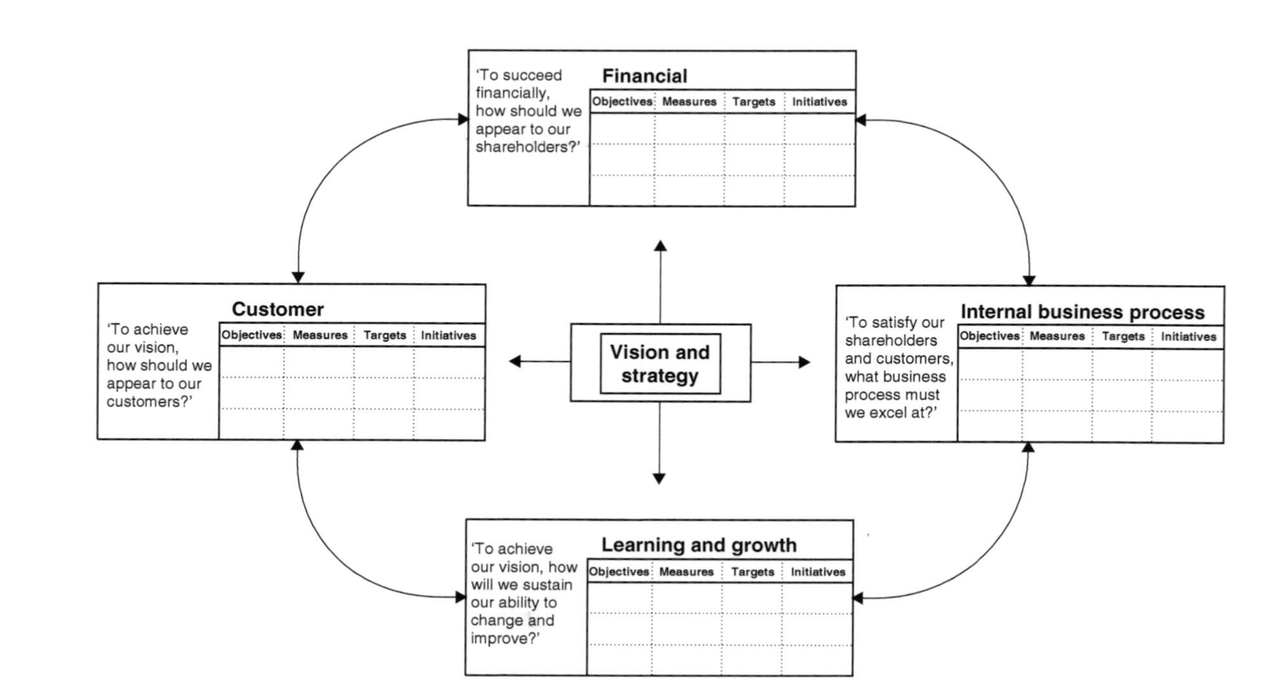

Figure 10.1 Kaplan and Norton's BSC (*Source*: Kaplan and Norton, California Management Review, 1996).

The published case studies by Kaplan (1996) provide examples of the application of the BSC in three areas: Chemical Bank, Mobil Corporation's US Marketing and Refining Division and United Way, a non-profit making community service based in Rhode Island, United States.

The application of the BSC has transformed methods of measuring a company's performance by financial indices alone. A recent publication by Basu (2001) has emphasised the impact of new measures on the collaborative supply chain. The Internet enabled supply chain or e-supply chain has extended the linear flow of supply chain to collaborative management supported by supplier partnerships. This has triggered the emergence of new measures especially in five areas:

- External focus
- Power to the consumer
- Value based competition and Customer relationship management
- Network performance and Supplier partnership
- Intellectual capacity

The design features and application requirements of the BSC are adapting to the collaborative culture of the integrated supply chain.

Basic steps

As shown in Figure 10.2, Kaplan and Norton (1996) recommend an eight step approach to introduce a BSC within an organisation.

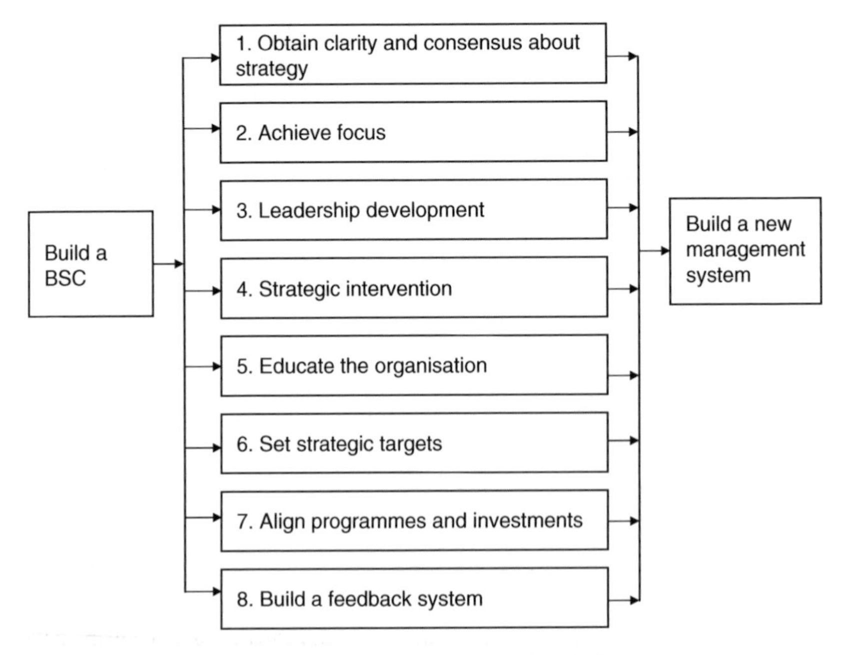

Figure 10.2 Kaplan and Norton's steps for implementation.

Our experience suggests that the comprehensive approach of Kaplan and Norton is underpinned by three fundamental criteria leading to the success of a performance management system including the BSC. These are:

1. Rigour in purpose
2. Rigour in measurement
3. Rigour in application

1. *Rigour in purpose*: Depending on the business objective, the metrics would vary in different industries. For example, in a pharmaceutical company, order fulfilment and compliance with regulatory standards are critical while in a bulk chemical industry asset utilisation may be more important. The metrics should be derived in alignment with company objectives and an emerging area for the four inter-linked perspectives of the BSC. The metrics are clearly defined, validated and accepted by users during a pilot exercise.
2. *Rigour in measurement*: The success of established metrics will depend on the effectiveness of data collection and monitoring systems. This could vary from a manual process on a spreadsheet to a sophisticated data warehouse. Table 10.1 shows examples of monitoring systems depending on their application.

Table 10.1 Examples of monitoring systems

Technology	Tools	Application
Local system	ERP Excel	Local sites
Visual factory	Manual Multimedia	Local system
Global system	ERP/SCM Internet Data warehouse	Local sites Regional Corporate

3. *Rigour in application*: The value of a well-designed and monitored BSC will be lost if the data is not used to improve and sustain performance. A process (such as Sales and Operation Planning) should be in place to review continuously the metrics and take action for performance improvement. Each measure should have a target both for the current year and the 'best in class' for the future. The measures are likely to be modified or reaffirmed to reflect the active usage of the BSC.

Worked-out example

The following example is based upon the BSC developed by the Worldwide Manufacturing and Supply (WM&S) Division of GlaxoWellcome Plc. in 1999, before the merger of the company with SmithKline Beecham to form GlaxoSmithKline (GSK).

Four overall measures and targets were established to meet the four primary objectives of W&MS as shown in Table 10.2.

Table 10.2 Manufacturing and Supply objectives

1. To support secure source of supply	1. Perfect order from suppliers
2. To support compliant and regulatory standards	2. Cost of poor quality
3. To support delivered point of sale	3. Perfect order to customers
4. To support 'best in class' cost	4. Cost of sales

For other supporting areas which are fundamental to the above objectives, additional measures were included in three perspectives:

1. Flexibility and Adaptability
2. Growth and Innovation
3. Health, Safety and Environment

The seven 'vital few' measures, described above and shown in Figure 10.3, sit at the top and they are calculated from sixteen component measures. For example, Perfect Order from Suppliers is derived from three components (viz. Order Fill, Vendor Managed Inventory and Quality Acceptance Rating) and the number of items.

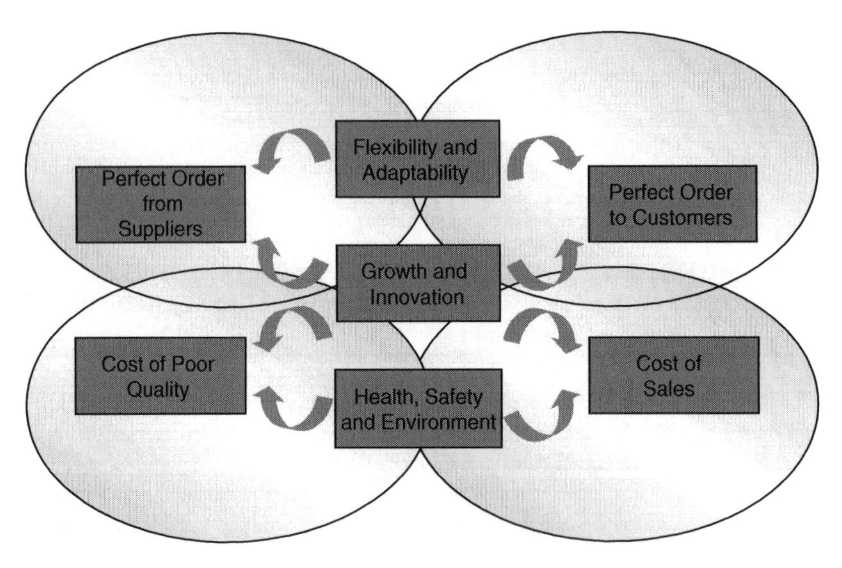

Figure 10.3 Glaxo Wellcome – Seven vital few (*Source*: GSK).

As shown in Figure 10.4, there are, in addition, 33 supporting measures which do not directly feed into the seven vital few, but which provide

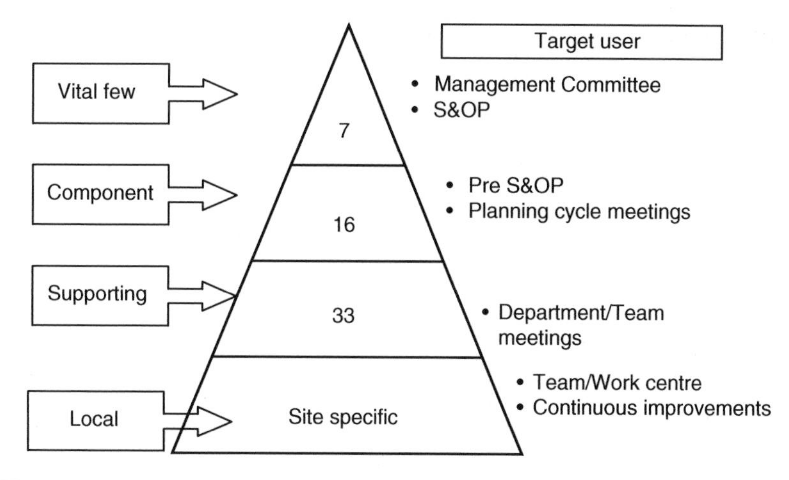

Figure 10.4 BSC measures and hierarchy.

important performance information for departmental reviews. There are also site specific measures which were not part of the BSC.

The above 56 measures were validated by a pilot exercise in three sites before the scorecard was rolled out. Following a continuous review of the benefits of the measures, the total number of measures was reduced to 27 metrics across the network.

Benefits

1. The BSC provides a sound basis to transform the approach of performance measurement to a management system.
2. It provides a tool for senior management to translate company strategies into longer-term goals.
3. It facilitates the cascading of high level measures to component and supporting measures and thus to explore the root causes of variations.
4. The key measures are balanced covering all perspectives of the business and also for both strategic and tactical purposes.
5. It provides a framework to implement a customised performance management system and culture.

Pitfalls

1. If it is not properly administered it could promote a league table culture of number games.
2. Sometimes well-intentioned measures could remain unexplained and could be seen as a corporate scheme for 'Big Brother'.
3. A major challenge for the BSC is too many measures lead to bureaucracy and too few measures do not fit all.

Training requirements

It is essential that the implementation of a BSC in an organisation is supported by training workshops. The workshop should contain the basic principles of the scorecard and the clarity and common definition of metrics. The need for local variation should also be considered seriously. The duration of such a workshop is usually 1 full day. The sharing of experience by visiting other sites is also part of the training requirements. There is a need for continuous review of requirements and metrics over the passage of time.

Final thoughts

The BSC is a powerful and effective technique for both large and medium-sized businesses in all sectors. However its effectiveness to a small enterprise of, say, less than 50 people should be reviewed carefully before its implementation.

R3: European Foundation of Quality Management

Background

The origin of the European Foundation of Quality Management (EFQM) relates particularly to the Malcolm Baldridge Award and also to the Deming Prize. The Malcolm Baldridge National Quality Award (MBNQA) has been presented annually, since 1988, to recognise companies in the United States who have excelled in quality management. The MBNQA criteria were based on seven categories, such as:

1. Leadership
2. Strategic Planning
3. Customer and Market Focus
4. Information and Analysis
5. Human Resource Focus
6. Process Management
7. Business Results

The Deming Prize was awarded mainly in Japan during the 1950s and 1960s based upon 10 examination viewpoints.

The EFQM was founded in the late 1980s by 14 large European companies to match the assessment criteria in Europe and the EFQM Excellence Model was launched in 1991. The model was regularly reviewed and an updated model was launched in 1999 which also included the RADAR (Results, Approaches, Deploy, Assess and Review) logic.

Definition

The EFQM excellence model is a framework for assessing business excellence and serves to provide a stimulus to companies and individuals to develop quality improvement initiatives and demonstrate sustainable superior performance in all aspects of the business.

As shown in Figure 10.5 the model is structured around 9 criteria and 32 sub criteria with a fixed allocation of points or percentages as shown in Table 10.3.

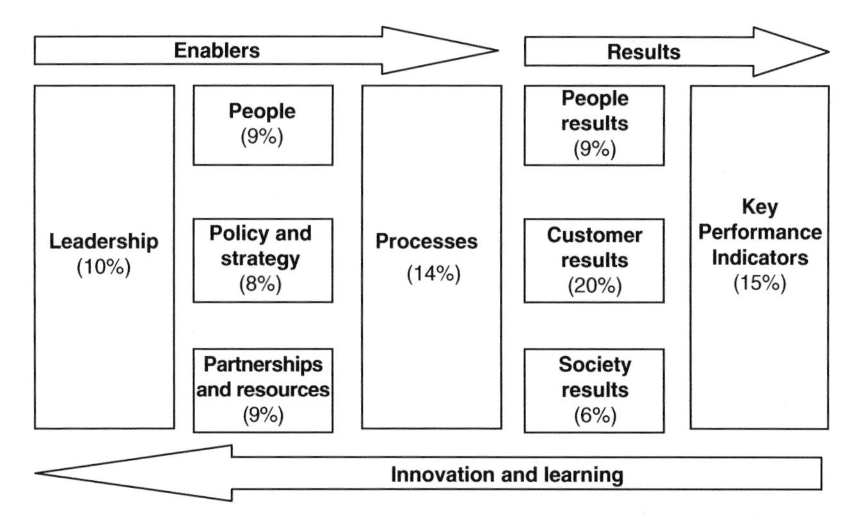

Figure 10.5 EFQM excellence model (for more details contact British Quality Foundation, London).

Table 10.3 EFQM criteria and points

Criteria	Points	No. of sub-criteria	%
1. Leadership	100	4	10
2. People	90	5	9
3. Policy and strategy	80	5	8
4. Partnership and resources	90	5	9
5. Processes	140	5	14
6. People results	90	2	9
7. Customer results	200	2	20
8. Society results	60	2	6
9. Key performance results	150	2	15
	1000	32	100

The criteria are grouped into two broad areas:

1. *Enablers*: How we do things – the first five criteria.
2. *Results*: What we measure, target and achieve – the second four criteria.

The sub-criteria (numbered as a, b or c) in an Enabler criterion have equal weighting. However, the weightings of Results criteria vary as

.6a, 7a and 8b – 75%
.6b, 7b and 8a – 25%

The scoring of each sub-criterion is guided by the RADAR logic which consists of four elements:

- Results
- Approach
- Deployment
- Assessment and review

The above elements and their attributes are applied to Enablers and Results as shown in Table 10.4.

Table 10.4 RADAR attributes

Elements	Attributes	Applies to
Results	Trends, targets, comparisons, causes, scope	Results
Approach	Sound, integrated	Enablers
Deployment	Implemented, systematic	Enablers
Assessment and review	Measurement, learning, improvement	Enablers

The words on the RADAR scoring matrix reflect the grade of excellence for each attribute and what the Assessor will be looking for in an organisation.

Application

The EFQM excellence model is intended to assist European managers to better understand best practices and support them in quality management programmes. The EFQM currently has 19 national partner organisations in Europe and the British Quality Foundation is such an organisation in the United Kingdom. Over 20 000 companies, including 60% of the top 25 companies in Europe, are members of the EFQM.

The model has been used for several purposes, of which the four main ones are given below:

1. *Self-assessment*: The holistic and structural framework of the model helps to identify the strengths and areas for improvement in any organisation and then to develop focused improvements.
2. *Benchmarking*: Undertaking the assessment of defined criteria against the model, the performance of an organisation is compared with that of others.
3. *Excellence Awards*: A company with a robust quality programme can apply for a European Quality Award (EQA) to demonstrate excellence in all nine criteria of the model. Although only one EQA is made each year for company, public sector and SME (small and medium enterprise),

several EQAs are awarded to companies who demonstrate superiority according to the EFQM excellence model.

4. *Strategy formulation*: The criteria and sub-criteria of the model has been used by many companies to formulate their business strategy.

A survey of EFQM members in 2000 showed a high proportion of the usage of the model in self-assessment and strategy formulation (see Figure 10.6).

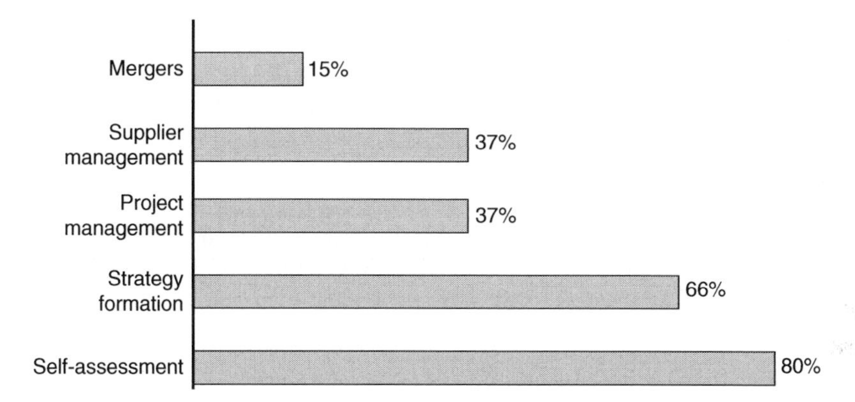

Figure 10.6 The use of EFQM (*Source*: British Quality Foundation; © Ron Basu).

The use of the model originated in larger business; however, the applications and interest have been growing amongst the public sector and smaller organisations. To satisfy these needs, special versions of the model are available for public sector organisations and SMEs.

Basic steps

The organisations can use the EFQM model without the involvement of a third party, except during the training period. The sequence of steps could vary depending on the approach adopted by a company. The following steps are recommended with an emphasis on self-assessment.

1. *Develop top management commitment*: It is essential to gain top management support, otherwise the process will have a limited value. Educate senior managers with the benefits and resource requirements for the self-assessment process.
2. *Plan a self-assessment process*: Select the appropriate self-assessment approach and the Excellence Model. If the company is not aiming for an EQA award then the criteria could be customise to companies VMOST (Vision, Mission, Objective, Strategy and Tactics). A roll-out plan is required starting from a pilot site.

3. *Select assessors and training*: Select a team comprising people with analytical and objective skills to carry out self-assessment. The team should be trained by EFQM approved trainers.
4. *Refine and communicate the self-assessment plan*: It is important that the right message of self-assessment (that it is not another audit from the centre) is communicated to all sites. It is not an academic exercise by a third party, but it is a process of continuous health check.
5. *Conduct self-assessment*: There are a number of ways of conducting self-assessment. A proven process is assessment of all criteria over 2 days by a team comprising both trained assessors (see Step 3) and local members.
6. *Develop action plan*: Based on the assessment in Step 5, identify the improvement opportunities and agree responsibilities and target dates for completion. Separate the activities which may required further investigation or investment.
7. *Implement action plan*: Set up an organisation to monitor progress and milestones. It is essential to secure senior management commitment.
8. *Repeat and review*: The self-assessment is a continuous process to achieve operational excellence. Steps 5, 6 and 7 should be repeated at least once a year.
9. *Consider certification or awards*: The company could consider an EQA award if the pure play EFQM Excellence Model is used. For a customised model the sites could be considered for their own company awards. For an EQA award, a report of up to 75 typed A4 pages is submitted to an EFQM Partner Organisation (e.g. BQF) and then the company is audited by external assessors.

Worked-out example

The following example illustrates the scoring method of the EFQM model.

Consider an organisation, following an assessment based on its performance, by adopting the RADAR logic, scored percentage values for each of the 32 sub-criteria. The scoring summary sheet in Table 10.5 shows that an organisation was awarded 531 points out of 1000 maximum points.

Benefits

1. It provides a holistic and realistic assessment of how good all the components and business processes of an organisation are.
2. It supports a balanced approach of assessing both qualitative (e.g. enablers) and quantitative (e.g. results) criteria covered in the model.
3. It provides a common language of enablers, results, assessment, scoring logic and certification for all types and sizes of organisation.
4. It brings the quality improvement initiatives into a single framework and creates a balance between different stakeholders groups.
5. The broad framework of the model allows its adaptation to a self-assessment checklist customised to individual company requirements.

Table 10.5 Summary scoring sheet

1. *Enablers criteria*											
Criteria number	1	%	2	%	3	%	4	%	5	%
Sub-criterion	1a	65	2a	55	3a	60	4a	50	5a	55
Sub-criterion	1b	55	2b	45	3b	65	4b	70	5b	45
Sub-criterion	1c	70	2c	60	3c	45	4c	50	5c	60
Sub-criterion	1d	65	2d	75	3d	60	4d	45	5d	50
			2e	65	3e	55	4e	55	5e	75
Sum		255/4		300/5		285/5		270/5		285/5
Score awarded		64		60		57		54		57

2. *Results criteria*								
Criteria number	6			%	7			%
Sub-criterion	6a	50	× .75	= 37	7a	60	× .75	= 45
Sub-criterion	6b	45	× .25	= 12	7b	50	× .25	= 13
Scores awarded				49				58
Criteria number	8			%	9			
Sub-criterion	8a	55	× .25	= 14	9a	70	× .5	= 35
Sub-criterion	8b	55	× .75	41	9b	40	× .5	= 20
Scores awarded				55				55

3. *Calculation of total points*

Criterion	Scores	Factor	Points awarded
1. Leadership	64	× 1.0	64
2. Policy and strategy	60	× 0.8	48
3. People	57	× 0.9	51
4. Partnership and resources	54	× 0.9	49
5. Processes	57	× 1.4	79
6. Customer Results	49	× 2.0	98
7. People results	58	× 0.9	52
8. Society results	55	× 0.6	33
9. Key performance results	55	× 1.5	83
Total points awarded			531

Source: EFQM, 1999

Pitfalls

1. The EQA award process requires a detailed report of up to 75 pages which is often viewed as a resource intensive and bureaucratic process.
2. The generic nature of the model, unless it is moderated by a well-framed assessor, has a risk of misinterpretation.
3. The allocated weighting factors of percentages (e.g. 20% for customers) do not necessarily reflect the relative priority and mission of a business.

4. The European level of EFQM often faces conflict, in a multi-national organisation, with MBNQA and ISO 9000.

Training requirements

The success of EFQM as a driver of a quality initiative depends on two key factors, viz. top management commitment and properly trained assessors. It is vital that an organisation has a team of assessors (say 1% of the workforce) trained and licensed by an EFQM approved organisation (e.g. British Quality Foundation). The senior managers and improvement team members should also receive 1-day awareness training by the company's own assessors.

Final thoughts

EFQM or its adaptation to a self-assessment process is an essential technique for achieving and sustaining operational excellence. However, an organisation has to be at an advanced stage of its quality programme to be able to use self-assessment in an effective manner.

R4: Sales and Operations Planning

Background

The classical concept of Sales and Operations Planning (S&OP) is rooted to the MRPII (Manufacturing Resource Planning) process. In the basic S&OP, the company operating plan (comprising sales forecast, production plan, inventory plan and shipments) is updated on a regular monthly basis by the senior management of a manufacturing organisation. The virtues, application and training of the S&OP have been promoted by Oliver Wight Associates (see Ling and Goddard, 1988) since the early 1970s.

In recent years the pace of change in technology and marketplace dynamics have been so rapid that the traditional methodology of monitoring the actual performance against pre-determined budgets set at the beginning of the year may no longer be valid. It is fundamental that businesses are managed on current conditions and up-to-date assumptions. There is also a vital need to establish an effective communication link, both horizontally across functional divisions and vertically across the management hierarchy to share common data and decision processes. Thus S&OP has moved beyond the operations planning at the aggregate level to a multi-functional senior management review process.

Definition

The traditional S&OP is a senior management review process of establishing the operational plan and other key activities of the business to best satisfy the current levels of sales forecasts according to the delivery capacity of the business.

Ling and Goddard (1988) summarise a 'capsule description of the process':

'It starts with the sales and marketing departments comparing actual demand to the sales plan, assessing the marketplace potential and projecting future demand. The updated demand plan is then communicated to the manufacturing, engineering and finance departments, which offer to support it. Any difficulties in supporting the sales plan are worked out … with a formal meeting chaired by the general manager'.

The outcome of the process is the updated operation plan over 18 months or 2 years (the 'planning horizon') with a firm commitment for at least 1 month.

The process is data driven. A report for each product family is prepared for the planning horizon and it is usually divided into up to five sections containing 'a single set of numbers' for Sales Plan, Production Plan, Inventory, Backlog and Shipments.

Application

S&OP has become an established companywide business planning process in the Oliver Wight MRPII methodology (see Wallace, 1990). It is now also known as Integrated Business Management or Senior Management Review.

The process has been developed and applied primarily for manufacturing organisations. The key members of all departments, such as R&D, Marketing, Sales, Logistics, Purchasing, Human Resources, Finance and Production, participate in the process but not in the same meeting. S&OP addresses the operations plan that deals with Sales, Production, Inventory and Backlog and thus is it expressed in units of measurements such as tons, pieces etc. rather than dollars or euros. The operation plan is reconciled with the business plans or budgets which are expressed in terms of money.

The S&OP or Senior Management Review process has been proven to be a key contributor to sustaining the performance level achieved through a Total Quality Management (TQM) or Six Sigma programme (Basu and Wright, 2003, p. 97). The S&OP agenda, in addition to its main focus of establishing the operation plan, contains the reviews related to performance and key initiatives. This provides an effective platform for senior managers of all functions to assess the current performance and steer the future direction of the business.

With appropriate adjustments for the units of the products, the S&OP process can also be applied to service industries. This will encourage the mangers in non-manufacturing sectors to review the demand, capacity, inventory and scheduling and enhance the synergy of different functions.

Basic steps

The diagram in Figure 10.7 shows the five steps in the S&OP process that will usually be present and the process can be adapted to specific organisation requirements.

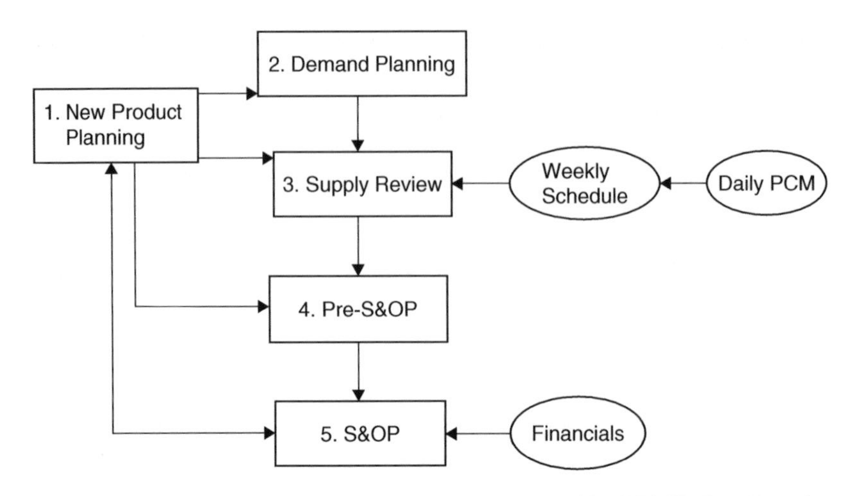

Figure 10.7 Senior Management Review Process (S&OP) (© Ron Basu).

New Product Review (Step 1): Many companies follow parallel projects related to the new products in R&D, Marketing and Operations. The purpose of this review process in Step 1 is to review the different objectives of various departments at the beginning of the month and resolve new product related assumptions and issues. The issues raised will impact upon the demand plan and the supply chain at a later stage of the process.

Demand Review (Step 2): Demand planning is more of a consensus art than a forecasting science. Demand may change from month to month depending on market intelligence, customer confidence, exchange rates, promotions, product availability and many other internal and external factors. This review at the end of the first week of the month, between Marketing, Sales, IT and Logistics, establishes agreement and accountability for the latest demand plan identifying changes and issues arising.

Supply Review (Step 3): In the current climate of increasing outsourcing and supply partnership, the capacity of supply is highly variable and there is a need to ensure the availability and optimisation of supply every month. This review, usually on the second week of the month, between Logistics, Purchasing and Production, establishes the production and procurement plans and raises capacity, inventory and scheduling issues.

Reconciliation Review (Step 4): Issues would have been identified in previous reviews of new products, demand and supply. The reconciliation step goes beyond the balancing of numbers to assess the business advantage and risk for each area of conflict. This review looks at issues from the business point of view rather than departmental objectives. This is also known as the Pre-S&OP Review and its aim is to minimise issues for the final S&OP stage.

Senior Management Review (Step 5): Senior Mangers or Board Members, with an MD or CEO in Chair, will approve the plan that will provide clear visibility for a single set of members driving the total business forward. The agenda includes the review of KPIs, business trends of operational and financial performance, issues arising from previous reviews and corporate initiatives. This is a powerful forum to adjust business direction and priorities. This is also known as the Sales and Operations Planning (S&OP) Review.

In each process step the reviews must address a planning horizon of 18–24 months in order to make a decision for both operational and strategic objectives. There may be a perceived view that S&OP is a process of aggregate/volume planning for supply chain. However, it is also a top level forum to provide a link between business plan and strategy.

Worked-out example

Table 10.6 shows a worked-out example of a product pack: Aquatic 500 in the unit of packs. The report is divided into four sections:

- Sales
- Stock
- Quality assurance (QA) release
- Production

In addition it contains some useful data, such as Production Batch Size (180), Lead Time (2 months) and Stock Target (3 months). The columns to the left of the line 'Today' show historical data and to the right is the information for the planning horizon in the future.

The data for the sales budget are taken from the annual business plan. 'Latest Forecast' represents what the sales and marketing teams are projecting based on the latest information. This data is updated every month. The stock target for each month is based on the sales forecast for the next 3 months, as the target is 3 months' stock cover. Due to a technical problem, production was suspended for 6 months and is resumed from this month. Therefore a backlog of order or negative stock (–1194) has been built up in the current month.

The projected stock is calculated by using the formula:

$$\text{Stock this month} = \text{Stock last months} - \text{Sales} + \text{Delivery}$$

For example, the projected stock for October $= -2094 - 500 + 3600 = 1006$

It is important to note that planned production should be in multiples of the batch size, i.e. 180 and the volume is available after 2 months' lead time.

Benefits

1. S&OP provides a practical up-to-date review of the operational plan of an organisation while meeting the business objectives of profitability, productivity and customer service.
2. It allows an excellent forum of senior managers of all functions to enhance the synergy to a common objective. The 'finger pointing' culture is thus eliminated.

Table 10.6 Sales and operations planning

Product toothpaste

Pack: Aquatic 500

Shelf life: three years

Unit: Packs

Supply source: Factory A

Prodn batch size: 150

Prodn + QA Lead time: Two months

Stock target: Three months

Month	−6	−5	−4	−3	−2	−1	1	2	3	4	5	6	7	8	9	10	11	12
	Jul	Aug	Sep	Oct	Nov	Dec	Jan	Feb	Mar	Apr	May	Jun	Jul	Aug	Sep	Oct	Nov	Dec
Sales																		
Budget	550	530	500	450	450	450	450	500	500	500	525	600	600	600	550	550	525	525
Latest forecast	450	550	350	500	480	450	450	500	400	450	400	500	400	500	400	500	400	500
Actual	375	610	465	395	413	484												
Performance (%)	83	111	133	79	86	108												
Stock																		
Target	1330	1250	1200	1250	1350	1450	1350	1250	1350	1300	1400	1300	1400	1300	1400			
Projected	1325	1400	1467	1483	1325	1306	1456	956	1456	1306	1506	1306	1506	1306	1506			
Actual	1430	1467	1433	1275	1256	1206												
Performance (%)	108	105	98	86	95	92												

QA release															
Planned	600	450	450	450	450	450	700	0	900	300	600	300	600	300	600
Actual	600	450	450	300	450	450									
Performance (%)	100	100	100	67	100	100									
Production start															
Planned	450	450	450	450	600	450	900	300	600	300	600	300	600		
Actual	450	450	450	450	700	0									

Year-to-date:	Sales budget:	2930	Actual sales:	2742	Performance versus budget:	94%
	Sales forecast:	2780			Performance versus forecast:	99%
			AV sales qty. (six months):	457		

3. It is data driven and based on a 'single set of numbers' for all departments and thus helps to reconcile disputes and planning issues.
4. It can play an effective role in sustaining the high level of performance achieved by a TQM or Six Sigma related programme.

Pitfalls

1. If the S&OP process is introduced without proper training, the managers may be obstructive for fear of detail and lack of understanding of the process.
2. A critical success factor is that it must be supported and chaired by the CEO or General Manager who can take a balanced approach related to past performance and future strategy and the degree of detail required.
3. The process will have limited value if all functions, especially Marketing, Sales, Logistics and Operations, are not involved at the appropriate stages of the business planning process.

Training requirements

All key managers, including the CEO or General Manager, should participate in a 2-day workshop on S&OP. It will also be beneficial for managers to sit in the S&OP meetings of other organisations where the process is fully operational.

There is a learning curve involved with S&OP. The first few meetings usually do not go well, but the process can become effective after 3 months. It is important to start S&OP as soon as possible.

Final thoughts

S&OP is an excellent data driven but people based holistic process to establish and update business plans and sustain business performance in keeping with the changes in the company and its marketplace. It is essential that it is underpinned by good training and led by the General Manager.

R5: Kanban

Background

The Toyota Motor Company of Japan pioneered the Kanban technique in the 1980s. As part of Lean Manufacturing concepts Kanban was promoted as one of the primary tools of Just-in-Time (JIT) concepts by both Tauchi Ohno (1988) and Shingo (1988). Inspired by this technique, American supermarkets in particular replenished shelves as they were emptied and thus reduced the number of storage spaces and inventory levels. With a varied degree of success outside Japan, Kanban has been applied to maintain an orderly flow of goods, materials and information throughout the entire operation.

Definition

Kanban literally means 'card'. It is usually a printed card in a transparent plastic cover that contains specific information regarding part number and quantity. It is a means of pulling parts and products through the manufacturing or logistics sequence as needed. It is therefore sometimes referred to as the 'pull system'. The variants of the Kanban system utilise other markers such as light, electronic signals, voice command or even hand signals.

Application

Following the Japanese examples, Kanban is accepted as a way of maximising efficiency by reducing both cost and inventory.

The key components of a Kanban system are:

- Kanban cards
- Standard containers or bins
- Workstations, usually a machine or a worktable
- Input and output areas

The input and output areas exist side by side for each workstation on the shop floor. The Kanban cards are attached to standard containers. These cards are used to withdraw additional parts from the preceding workstation to replace the ones that are used. When a full container reaches the last downstream workstation, the card is switched to an empty container. This empty container and the card are then sent to the first workstation signalling that more parts are needed for its operation.

A Kanban system may use either a single card or a two cards (move and production) system. The dual card system works well in a high up-time process for simpler products with well-trained operators. A single card system is more appropriate in a batch process with a higher changeover time and has the advantage of being simpler to operate. The single card system is also known as 'Withdrawal Kanban' and the dual card system is sometimes called 'Production Kanban'.

The system has been modified in many applications and in some facilities although it is known as a Kanban system, the card itself does not exist. In some cases the empty position on the input or output areas is sufficient to indicate that the next container is needed.

Basic steps

1. Select the operation for the Kanban system and decide whether a single or dual card system will be applied. We recommend the single card system for its simplicity.
2. Determine the number of Kanban containers to set the amount of authorised inventory. Use the following formula:

$$\text{Number of containers} = \frac{\text{Demand in lead time} + \text{Safety stock}}{\text{Size of container}}$$

3. Design and procure the standardised containers and Kanban cards.
4. Develop and implement the workstation layout. Carry out a pilot run.
5. Train operators and activate the Kanban system by following some basic rules:
   - Each container must have a Kanban card.
   - Each container must contain the exact quantity stated on the card.
   - The containers are pulled only when needed by the next downstream station.
   - No defective parts are sent.
6. Review the process regularly and aim to reduce the number of Kanbans and the time period.

Worked-out example

The following example is based upon the experience of Level Industrial, the Brazil subsidiary of Unilever in Sao Paulo.

Lever Industrial was engaged in the batch production of industrial detergents comprising nearly 300 stock keeping units which varied from a 500 kg draw to a 200 gm bottle. After carrying out a Pareto analysis the team selected three fast-moving products for a pilot Kanban system. These products in total accounted for 18% of output.

The company adopted for each product, a simple single card Kanban system consisting of five stages as shown below (Figure 10.8).

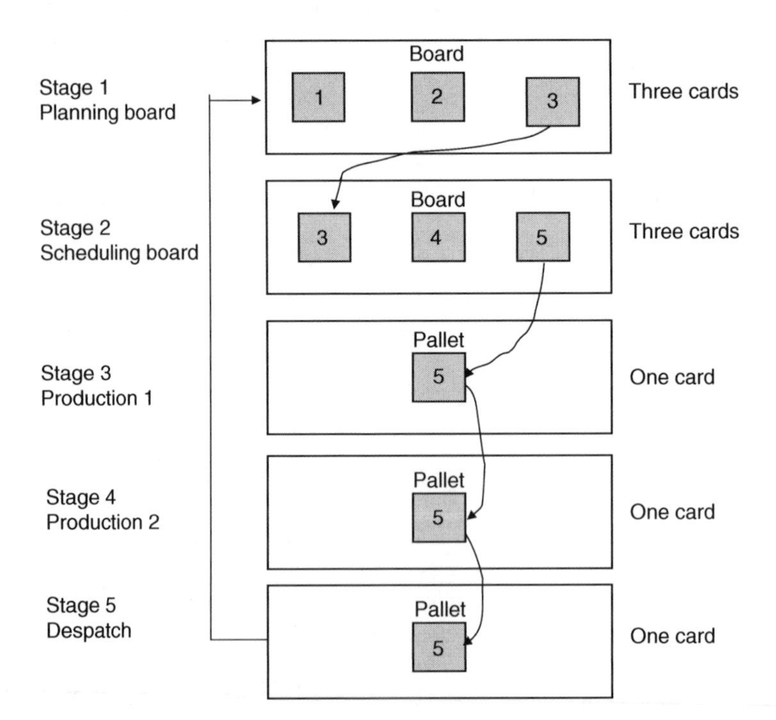

Figure 10.8 Kanban system (© Ron Basu).

Both the planning board and the scheduling board contain three cards each as a buffer between the variability of production cycle time and the availability of materials.

When the card arrives from the despatch (Stage 5) it is kept on the planning board and planning for the product starts. When the planning board is full with three cards, the third card is passed to the scheduling board and production scheduling is ensured. Similarly when the scheduling board is full, the third card is transferred to the pallet at the Production Station 1 and actual production begins.

When the pallet in Stage 3 (Production 1) is full, the card then moves to the next station (Production 2) in Stage 4, and then on to despatch in Stage 5. After the goods are despatched, the card returns to the planning board and the next cycle begins.

The pilot exercise was successful. It achieved an improvement in customer service which rose from 84% to an excellent 98% and inventory was also reduced. The Kanban system was extended to nine additional key products. The manual system was retained for the above five stages, although both the planning and stock adjustment processes were supported by MFG-Pro, the Enterprise Resource Planning (ERP) system.

Benefits

1. Kanban enables only small inventory through the plant and pulling only when needed thus allowing only a small quantity of faulty or delayed material.
2. Kanban minimises the negative aspects of inventory management including obsolescence, occupied space, working capital, increased material handling and poor quality.
3. Kanban uses standardised containers conducive to efficient material handling and lower costs.
4. It aims to create work sites that can respond to changes quickly empowering the operators to exercise their initiatives.
5. It facilitates the re-engineering of the process and works in harmony with JIT techniques.

Pitfalls

1. It is an inflexible process, as the transfer batch is fixed. Therefore it can cause additional stoppage periods.
2. Kanban is inappropriate for high mix, slow mover variants. It struggles with cyclic or seasonal demand.
3. It is perceived as a low technology manual process and comes into conflict with the push MRPII/ERP systems. (However there are good examples of a computerised MRPII system working hand-in-hand with a Kanban call-off scheme).
4. The application is visible as a solution to a part of the total operation and often not appreciated by employees who are not directly involved with the Kanban system.

Training requirements

Although the fundamental principles of the Kanban technique are not complex, its training needs should not be underestimated. The two card system requires a team of well-trained and motivated workers.

The operators should be trained for 1 day in a classroom environment and this should be followed by 2 or 3 days of on-the-job training in the shop floor environment.

Final thoughts

The attractiveness of a Kanban system cannot be ignored even in an environment of flexibility and ERP systems. Select and apply Kanban for fast-moving products containing the repetitive manufacturing of discrete units in large volumes which can be held steady for a period of time.

R6: Activity Based Costing

Background

The classical method of costing allocates overhead costs to each product according to the amount of direct labour required to make that product. These days labour is a much smaller element of the product costs and thus the product costs are disproportionately distorted. Alternatively, the overhead cost is allocated according to the volume of the product. This could lead to increased consumer demand for undercosted low volume product and decreased consumer demand for overpriced high volume products.

The above problems associated with the traditional overhead allocation methods were first highlighted by Johnson and Kaplan (1987) who demonstrated that historical methods of cost accounting have led to cost distortions. These distortions of both inaccuracy and inappropriateness have harmful effects on product profitability. The cost of quality is also affected by the traditional accounting system. A certain Prof. Cooper in the United States suggested the name 'Activity Based Costing' or ABC.

ABC seeks to correct these distortions by assigning indirect costs to products and services by using appropriate activity drivers that reflect resource consumption by the cost objectives. This also enables the quality cost associated with an activity to be more easily obtained.

Definition

ABC is an accounting technique that allows an organisation to assign more accurate product or service costs by understanding the activities that create cost and allocating overheads on those activities. Overhead costs and their causes are analysed so that they can be transferred, wherever possible, into direct costs.

There are some standard terminology used in ABC. These include:

Activities: Activities are the types of work done and which consume resources in an organisation.
Cost Pool: A cost pool is the total amount of costs which may be derived from different departments, associated with an activity.
Cost Driver: A cost driver is an item that triggers an activity. Costs are assigned according to the number of occurrences associated with a cost driver.

The cost drivers are used to derive the cost of activities to cost pools. There is one cost driver for each cost pool.

The relationship between the key elements of budget at completion (BAC) is illustrated in Figure 10.9.

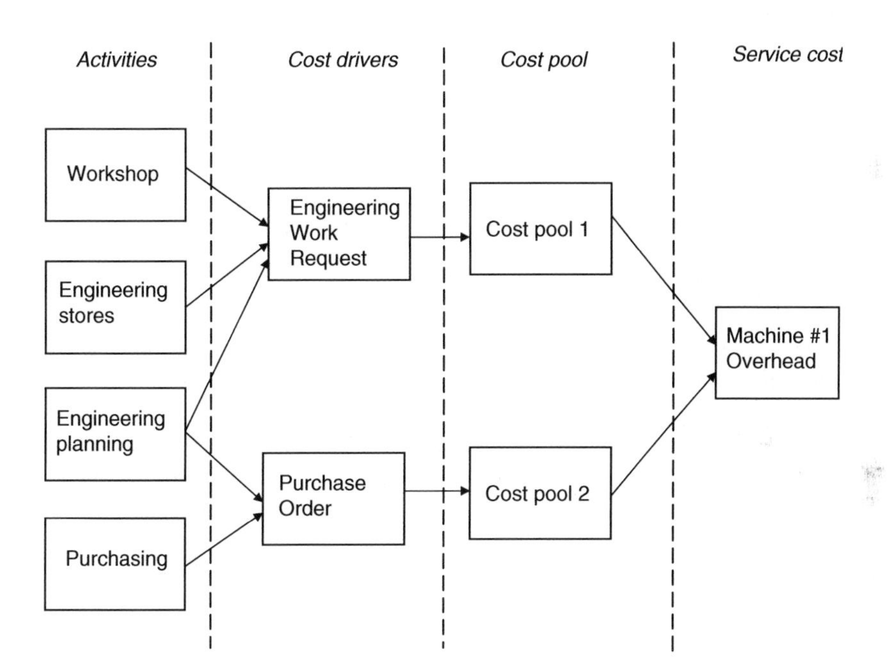

Figure 10.9 ABC terminology (© Ron Basu).

Application

ABC is based on the concept that products give rise to activities which drive costs. The examples of such cost drivers in the procurement activity, e.g. include purchase requisitions, quotations, purchase orders, invoices etc. In providing more accurate allocation of costs, ABC has also proved to be useful in identifying cost reduction opportunities.

Activity Based Management (ABM) applies ABC generated data for planning, controlling and cost-effectiveness. Since the early 1990s, many companies have attempted to apply ABC in their cost management initiatives. The UK companies with an active interest in ABC include British Airways, Unipart, Norwich Union and Cummins Engine.

A company is particularly suited to the application of ABC when:

- The company produces a wide variety of customised products
- The company has high overhead costs

It is not the answer to all costing problems, however. It is just not appropriate, e.g. for a business where simple products are produced. In fact it is now generally accepted that ABC can be used alongside a more traditional accounting system in the same company. The best approach has been to perform ABC for analysis purposes and retain the standard costing system for bookkeeping.

ABC can also be useful to a service organisation, provided that the activities are relatively homogenous and its output can be defined. Examples of such applications are hospital and insurance services.

Basic steps

The basic premise of ABC is that activities use resources and products consume activities. ABC uses many cost drivers. However, conventional methods typically use only one. Consequently it is expected that the ABC method will increase the overhead allocation accuracy.

The basic steps for developing an ABC system are:

1. Define and analyse activities: Activities are usually, but not necessarily, related to functional departments. However one department may contain several activities.
2. Determine the cost driver for each activity: Cost drivers may be measured in terms of volume of transactions undertaken. For example, for the quality control activity, the cost driver is given by number of inspections.
3. Identify cost pools: activity costs with the same cost driver are collected in activity pools.
4. Assign cost products or services: The costs of activities in cost pools are assigned to products or services based on cost drivers.

Worked-out example

The following example is taken from Maskell (1996), p. 107.

Consider a customer service department where two cost reports are made for comparison purposes, as shown in Table 10.7.

It is evident that expediting, correction and issuing credits are non-value added activities but they represent $520 000, nearly half of the total cost.

Benefits

1. ABC provides a means of increasing the accuracy of cost allocation for both manufacturing and service organisations.

Table 10.7 Traditional versus ABC reports

Traditional cost report	Activity based cost report
Salaries $920,000	Take orders $600,000
Space $100,000	Expedite orders $140,000
Depreciation $100,000	Correct orders $120,000
Supplier $60,000	Issue credits $160,000
Other $20,000	Amend orders $60,000
	Answer questions $40,000
	Supervise $80,000
Total $1,200,000	Total $1,200,000

2. It allows a better understanding of the cost of making a product and providing a service and thus helps to focus on priority activities for improvement.
3. ABC moves away form the notions of short-term fixed and variable cost and focuses on the variability of the cost in the longer term. Hence ABC has directed management attention from product costing to improving business processes leading to ABM.
4. The principles of ABC can be very effective in the analyses of customer profitability and cost of poor quality. Price estimates are also enhanced by ABC.

Pitfalls

1. ABC has its limitations in high technology industries and in allocating overheads where cause of these overheads are unknown.
2. The process of calculation in ABC is still viewed as complex. Empirical surveys in the United Kingdom (Bromwich and Blumani, 1989) showed that companies expressed a great reluctance to change from their traditional cost accounting systems.
3. The process is time consuming and its benefits are marginal in comparison to the effort associated with the detailed analyses required.

Training requirements

The use of the ABC technique should be restricted to qualified management accountants. However the project team members should benefit from attending the awareness workshop on ABC so that they can contributed to the identification of cost saving opportunities.

Final thoughts

ABC is conceptually simple and a useful technique for product pricing and profitability analysis. However, due to the complexity of calculation it should be used sparingly with support for qualified accountants.

R7: Quality Management Systems

Background

In this book, ISO 9000 represents the general area of accredited Quality Management Systems which relate to the organisation, procedures and processes for implementing quality management. In 1979 the British Standard Institute (BSI) issued the BS5750 series of quality management systems standards. BS 5750 became ISO 9000 when in 1994 the International Standard Organisation (ISO) created its now famous ISO 9000 by adoption of the BS 5750 together with parts of other national quality management standards. The ISO 9000 series of standards ran to around 20 different standards of which the main ones were:

ISO 9001	For design, development and production
ISO 9002	For production, installation and servicing
ISO 9003	For final inspection and test
ISO 9004	For quality management systems

In addition, ISO 14000 was also published for environmental management standards.

The accreditation to ISO 9000 became very popular in the 1990s with Government subsidies and customers asking their suppliers for confirmation of their accreditation. However, the 1994 version of ISO 9000 came into disrepute for four main reasons.

1. The numbering system left a lot to be desired. A company may be approved of ISO 9001, ISO 9002 or ISO 9003, but some customers still intended to audit them.
2. The Government (e.g. Department of Trade and Industry in the United Kingdom) certified a large team of consultants and provided subsidies to promote accreditation and thus the standard of assessment could not be regulated.
3. The emphasis being on the maintenance of written quality procedures it was viewed as 'institutionalising existing bad practice'.
4. The accreditation focused on one area of the organisation or process and therefore was not found to be a driver for improving the total business.

The vision for Phase 2 to develop a single quality management standard and address the above issues was conceived in 1996 and a new version is referred to as ISO 9001: 2000 (see Figure 10.10). In the following sections we have described and analysed the new version of ISO 9000.

Definition

ISO 9001: 2000 is the updated quality management system which specifies the requirements for an organisation to demonstrate it ability to provide products and processes that fulfil customer satisfaction.

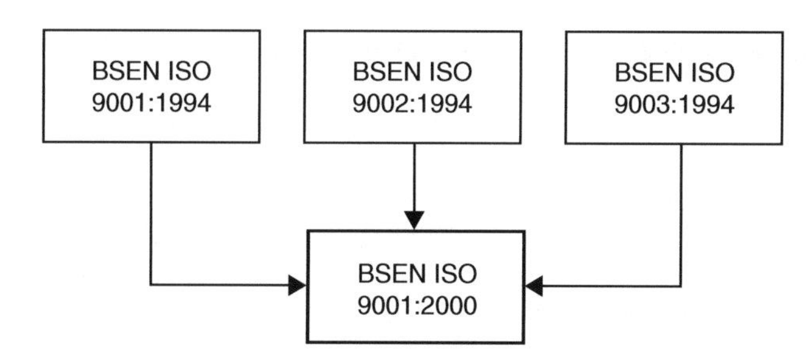

Figure 10.10 Link between ISO 9000: 1994 and ISO 9000: 2000 (© Ron Basu).

As shown in Figure 10.11, ISO 9001: 2000 contains significant changes from the 1994 standard and reflects the integration of six main areas.

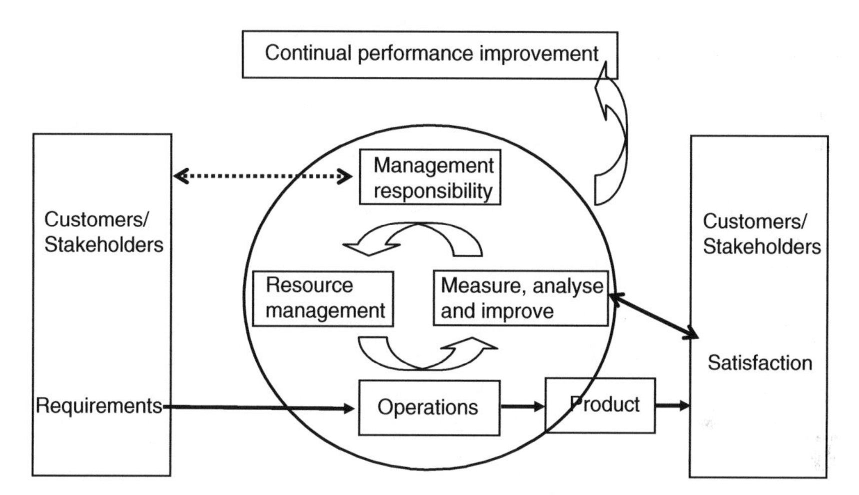

Figure 10.11 ISO 9000: 2000 Quality Model (for more details contact British Quality Foundation, London).

1. *Management responsibility*: More emphasis on senior management involvement.
2. *Resource Management*: Less emphasis on paperwork and more on resources and business processes.
3. *Product realisation*: Production and service under controlled conditions.
4. *Measurement, analysis and improvement*: Requires measurement of processes.
5. *Customer focus*: Requires measurement of customer satisfaction.
6. *Continuous improvement*: Focuses the continuous improvement of both the processes and quality management system.

The first four are the fundamental requirements to achieve number 5 (i.e. customer focus) and number 6 (i.e. continuous improvement).

Eight guiding principles were identified to meet the requirements of ISO 9001: 2000. These are:

- Role of leadership
- Involvement of people
- Systematic approach to management
- Customer focus
- Business process approach
- Factual approach to decision making
- Continual improvement
- Mutually beneficial supplier relationships
- New requirements

Application

The 1994 version of ISO 9000 was extensively applied in both SMEs and large organisations. Many organisations have been updated their accreditation to the new version of ISO 9001: 2000.

It is a long and expensive process to gain ISO 9000 accreditation. It has been used for two primary objectives:

- To gain the customer's acceptance as a preferred supplier.
- The quality system as a pillar in an organisation's approach to TQM.

However ISO 9000 registration is not a pre-requisite for TQM. Many organisations, particularly in Japan, achieved excellent quality standards without the support of ISO 9000. It is also indisputable that the development and maintenance of procedures and control, as required by ISO 9000, help to sustain quality standards. As shown in Figure 10.12, ISO 9000 can act as a stopper that prevents the quality standard going in reverse.

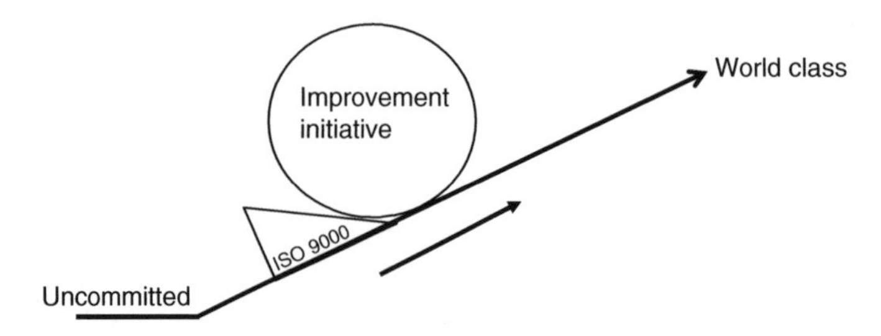

Figure 10.12 Quality improvement and ISO 9000 (*Source*: Dale 1999, p. 273; © Ron Basu).

The fragmented approach of the 1994 version still lingers on and appears to encourage the separation of a business into compartments where the requirements for ISO 9000 could be met. Organisations, without a TQM or Six Sigma programme, find it difficult to involved every function in a company-wide registration

Many organisations or people are still confused about the relationship between TQM and ISO 9000. They are not alternatives, but they complement each other. As mentioned earlier, there are examples of successful TQM programmes without ISO 9000 (Sayle, 1991). Some partial improvements driven through the 1994 version of ISO 9000 would not lead an organisation to TQM. The new version ISO 9001: 2000 has some potential to be an essential feature of a companywide quality programme.

Another area of confusion is the relationship between ISO 9000: 2000 and EFQM. If the new version of ISO 9000 is applied to the whole of the organisation then it relates closely to EFQM. They have key principles in common (except perhaps the focus on supplies is less visible in ISO 9000) and the possible differences are in the scope of application. It is arguable that the differences, albeit limited, are promoted by the governing bodies of these two closely related systems.

Basic steps

1. *Prepare for the quality management system*: The organisation should state clearly the purpose, scope and benefits of going for ISO 9000. A steering committee should be established, ideally headed by the CEO.
2. *Train a team*: A QMS team is formed for a large and medium-sized organisation. Train the team with the fundamentals of ISO 9000: 2000 so that the members can prepare the QMS document and conduct internal quality audits. Small organisations (of say less than 50 people) usually depend on external consultants.
3. *Prepare the QMS document*: The document for the quality management system is determined by the nature of the business, but follows the well-defined guidelines of ISO 9000: 2000. It should cover the following checklist:
   a. Management responsibility
      i. Customer requirements
      ii. Quality policy, objectives and planning
      iii. Quality manual
      iv. Management review
      v. Control procedures
   b. Resource Management
      i. Human resources
      ii. Business processes
   c. Product realisation
      i. Design and development
      ii. Procurement
      iii. Production and service delivery processes
      iv. Customer service
   d. Measurement, analysis and improvement
      i. Measurement and monitoring
      ii. Data analysis
      iii. Improvement
      iv. Control of non-conforming products

4. *Pre-audit and identify gaps*: It is important that an internal quality audit is conducted including the review of the QMS document by a qualified auditor. Involvement of the internal QMS team during this quality audit is also very important.
5. *Registration for ISO 9000*: When the QMS system is ready and supported by the pre-audit, a registration is sought in an accredited ISO 9000 certification body. The body will then supply an information pack and then the necessary terms are agreed with the accredited body.
6. *Certification*: The most appropriate time for assessment is decided after the QMS has been effectively running for 6 months. The assessment is carried out by a small team of independent assessors approved by the certification body. Any non-conformance with standards is rectified before a certification is awarded.
7. *Review*: The certification bodies usually follow a system of routine surveillance and revisit the organisation every 2 years at the invitation of the company.

Worked-out example

The following example has been adapted from a published case study by McLymont and Zuckerman (2001).

Silberline Manufacturing of Lansford, United States, completed a full transition to ISO 9001–2000 three years after registering to ISO 9001: 1994. The company is a global manufacturer of aluminium pigment and special products for coatings and the plastics industry.

The transition was led by a cross functional steering committee and lasted for 1 full year from April 2000.

While no major changes were needed in product realisation processes, the site was organised around process flows that began with sales and ended with products delivered to customers. The process was then weakened until it was in control.

Rewriting the quality manual was the most frustrating part of the transition. The team abandoned the 20 element model of ISO 9001: 1994 and based the new manual on the ISO 9001: 2000 process model.

An ERP was already in place. All departments met monthly to discuss a 'red, yellow, green light' report which included output generated by the ERP system. The report was similar to a BSC format. The transit audit was conducted by BSI Inc., the US operation of the British Standards Institution. In spite of all of these trials, Silburn described the changeover to the new ISO 9001: 2000 process as fairly seamless.

Benefits

1. The new version of ISO 900: 2000 reaches out to address companywide quality issues and ensures that specified customer requirements are met.

2. Like the EFQM excellence model it provides a framework of assessing a TQM or Six Sigma programme towards achieving operational excellence.
3. It has an international status covering all types of organisation across geographical and regional boundaries.
4. It establishes a discipline of improved controls and standard procedures, thus preventing the duplication and compromising of activities.
5. It provides an objective platform to enhance teamwork when it is applied as a companywide programme.

Pitfalls

1. It still suffers from the chequered history of the 1994 version and there are many 'agnostics' towards the system. It is being viewed as a bureaucratic process.
2. It is far from becoming a people process leading to self-assessments. The methodology is guarded by 'qualified' assessors.
3. The external assessors have often oversold the expectations of ISO 9000 and the certification has failed to deliver all the benefits.

Final thoughts

The new version of ISO 9000: 2000 can be used as an effective technique to improved process control and business performance in all types of organisation, whether large or small, manufacturing or service, private or the public sector. However it has a long way to go to establish its credibility to make it people friendly and to remove the artificial demarcation with excellence models like EFQM.

R8: Lean Thinking

Background

As with all facets of the quality movement the origin of Lean Thinking is in manufacturing. Lean Thinking philosophy, and make no mistake Lean is more than a system it is a philosophy, began with Japanese automobile manufacturing in the 1960s, and was popularised by Womack et al. (1990) in 'The Machine that Changed The World'. It is essentially the story of the Toyota way of manufacturing automobiles.

The Lean Thinking, sometimes referred to as Toyotaism or Toyota Production System, is that materials flow 'like water' from the supplier through the production process onto the customer with little if any stock of raw materials or components in warehouses, no buffer stocks of materials and part finished goods between stages of the manufacturing process, and no output stock of finished goods.

Definition

Womack and Jones (1996) proposed five Lean principles based on Toyota Production Systems, viz. Value, Value Stream, Flow, Pull and Perfection. Before anything can be eliminated it first has to be identified. The Toyota approach to identifying areas of waste is to classify waste into seven 'mudas'.

The seven 'mudas' are:

- Excess production
- Waiting
- Movement or transportation
- Unnecessary motion
- Non-essential process
- Inventory
- Defects

The approach is to identify waste, find the cause, eliminate the cause, make improvements and standardise (until further improvements are found).

Application

However the application of Lean principles has moved with time and experience of organisations in both manufacturing and service sectors. Until recently supply chains were understood primarily in terms planning the demand forecasts, upstream collaboration with suppliers and planning and scheduling the resources. Emphasis perhaps is shifted to provide what the customers want at a best in class cost. Cost reduction is often the key driver for lean, but it also about speed of delivery and quality of products and service. The competition for gaining and retaining customers and market share is between supply chains rather than other functions of companies. A supply chain therefore has to be lean with four inter-related key characteristics or objectives:

1. Elimination of waste
2. Smooth operation flow
3. High level of efficiency
4. Quality assurance

Elimination of waste

The lean methodology as laid out by Womack and Jones (1996) is sharply focussed on the identification and elimination of 'mudas' or waste and their first two principles (i.e. Value and Value Stream) are centred around the elimination of waste. Their motto has been, 'banish waste and create wealth in your organisation'. It starts with value stream mapping to identify value and then identify waste with Process Mapping of valued processes and then systematically eliminate them. This emphasis on waste elimination has probably made lean synonymous to absence of waste. Waste reduction is often a good place to start in the overall effort to create a lean supply chain because it can often be done with little or no capital investment.

The tools for the elimination of waste include Value steam mapping, Process mapping and Poka-Yoke and there tools have described earlier in this book.

Smooth operational flow

The well-publicised JIT approach is a key driver of Lean Supply Chain and, as we have indicated earlier, it requires materials and products flow 'like water' from the supplier through the production process onto the customer. The capacity bottlenecks are eliminated, the process times of work stations are balanced, and there is little buffer inventories between operations. Smooth operation flow requires the applications of appropriate approaches. Three of the most frequently applied approached are:

- Cellular manufacturing
- Kanban pull system
- Theory of constraints

Kanban system has been described as a stand alone technique earlier in this chapter. In cellular manufacturing concept traditional batch production area is transformed into flow line layouts so that ideally a single piece flows through the line at any a time. In practice an optimum batch size is calculated starting with the most critical work centres and the largest inventory carrying costs. Action is taken for improvement at the work centres and methods that have greatest impact on the throughput, customer satisfaction, operating cost and inventory carrying charges. The theory of constraints is a management philosophy developed by E M Goldratt (1992). It enables the managers of a system to achieve more of the goal that system is designed to produce. The concept or the objective is not new. However in service operations where it is often difficult to quantify the capacity constraint Total Productive Maintainence (TOC) could be very useful.

High level of efficiency

The more popular concepts of lean operations tend to be the concepts of muda, flow and pull system. A preliminary analysis of all these methods, as we have described earlier, however, highlights the fact that all assume sufficient machine availability exists as a pre-requisite. In our experience for many companies attempting a lean transformation this assumption is not true. Machine availability depends on maximising the machine up time by eliminating the root causes of down time. The ratio of up time and planned operation time is the efficiency of the operation. Therefore in order to make lean concepts work it is vital that the pre-condition of running the operations at a high level of efficiency should be met.

There are many tools and techniques used to achieve and sustain high level of efficiency and most useful and popular ones are TPM, OEE, SMED and 5Ss. These tools and techniques have been described earlier in this book.

Quality assurance

Womack and Jones (1996) propose Perfection as the fifth Lean principle and according to this a lean manufacturer sets his/her targets for perfection in an incremental (Kaizen) path. The idea of TQM also is to systematically and continuously remove the root causes of poor quality from the production processes so that the organisation as a whole and its products are moving towards perfection. This relentless pursuit of the perfect is key attitude of an organisation that is 'going for lean'.

This drive for quality assurance has now been extended beyond TQM to Six Sigma with additional rigour in training deployment (e.g. Black Belts and Green Belts), the methodology of DMAIC (e.g. Define, Measure, Analyse, Improve and Control), and measurement (both variances and savings). The principles of Six Sigma are embedded in the path towards perfection in a lean supply chain and Six Sigma has now moved to Lean Sigma and FIT SIGMA™. Basu and Wright (2003) explains that the predictable Six Sigma precisions combined with the speed and agility of Lean produces definitive solutions for better, faster and cheaper business processes.

Basic steps

The basic steps of Lean Thinking are centred around the almost sequential application of four key characteristics in the operation, process or department under investigation, viz. elimination of waste, smooth operation flow, high level of efficiency and quality assurance.

The steps for elimination of waste should include:

1. Analyse the purpose and concept of value and value stream mapping.
2. Identify the essential elements of value stream mapping.
3. Assess the appropriate mapping tools to create your value stream map.
4. Establish the steps for selecting the value stream to be mapped:
   a. Calculations
   b. Symbols
5. Pinpoint the criteria to determine value added and non-value added activities.
6. Develop the appropriate improvement strategies and ensure continuous improvement of the value stream process.
7. Create the future-state value stream map to determine what your value stream will look like after applying improvements.
8. Evaluate new opportunities to create new value streams.

Having identified and minimised waste efforts are focused on smooth operational flow and high level of efficiency. The principles of quality assurance are instilled to sustain the high level of efficacy in a lean operation.

Worked-out example

There are abundance of examples of Lean Thinking in both manufacturing and service industries. The following example illustrates how a simple application of lean principles helped to reduce the procurement cycle time in a pharmaceutical company.

Platinum catalyst is used for production of an Active Pharmaceutical Ingredient (API) in an Eastern European Pharmaceutical Company (henceforth referred as 'company'). Used catalyst is sent back to supplier who recovers Platinum and uses it for production of fresh catalysts. During that cycle certain quantity of catalyst evanesces and new quantity has to be purchased periodically to maintain required level of inventory. The catalyst is expensive because of Platinum and the related cost of capital for required catalyst inventory is significant. A task team led by a Six Sigma Black Belt was formed to reduce the cycle time of procuring the platinum catalyst.

For the monitored period the mean regeneration times depending on the supplier varied between 77 days and 69 days (Year 2003) and during year 2004 values were marginally better year before. During Year 2004 significantly better results were achieved also for the transport time and the average transport time was 5 days ±2 days what was acceptable. The biggest influence on overall cycle time was the regeneration of catalyst. This was clearly the supplier's responsibility and the company could not directly influence that process. The regeneration time specified in contract between the company and each supplier was 10 weeks for one mayor supplier and 11 weeks for another for year 2004. To test supplier's ability to fulfil new requirement the company asked each of supplier to deliver next shipments of regenerated catalyst till the end of year 2004 within 9 weeks instead 11 (including transport). One of the approved suppliers answered positively but asked for some adjustment in packaging of spent catalyst which did not require additional cost. That allowed the company not to buy new quantity of 1.000 kg of fresh Platinum catalyst and generated a saving in cost of capital of 20.000 USD in last 3 months of the year 2004.

As a result of these improvements cycle time was reduced by 30% and the inventory of the catalyst reduced from 7.728 kg to 4.500 kg. The overall annual savings related to avoidance of cost of capital needed for buying of new quantity of catalyst was $ 408 615 per annum.

Benefits

The benefits of Lean Thinking are both tangible or quantifiable and intangible or benefits in business culture and environment.

The tangible results will vary, but here are some typical savings and improvements:

Reduce

Lead time	50–90%
Floor space requirement	5–30%
Work-in-progress	60–80%

Increase

First-time yield	50–100%
Throughput	40–80%
Productivity	75–125%

The intangible and environmental benefits of Lean Thinking include:

- Continual improvement culture focused on uncovering and eliminating wastes, hidden wastes and waste-generating activities.
- Eliminates overproduction, thereby reducing waste and the use of energy and raw materials.
- Fewer defects from processing and product changeovers- reduces energy and resource needs; avoids waste.
- Less in-process and post-process inventory needed; avoids potential waste from damaged, spoiled, or deteriorated products.
- Less floor space needed; potential decrease in energy use and less need to construct new facilities.
- Quick, sustained results without significant capital investment.
- Environmental benefits are more broadly realized by introducing lean to existing suppliers rather than finding new, already lean suppliers.

Pitfalls

Many organisations encounter some of these typical pitfalls or barriers to Lean:

- Belief that 'our company is different', and that Lean does not really apply to 'our company'.
- Lack of long-term serious management commitment to Lean often results in a 'back to business as usual' syndrome.
- Lack of a clear link to the business' strategic goals. Firms take a piece-meal approach to implementing the principles of Lean and partial efforts often fail.
- Lack of employee education can cause Lean efforts to fail. The implementation of Lean requires a change of mind and heart of every employee.
- Believing Lean is *only* about manufacturing.

Final thoughts

Womac and Jones (1996) wrote, 'Lean thinking can be applied to any company anywhere in the world .but the full power of the system is only realised when it is applied to all element of the enterprise'. The application of Lean thinking has been successful when it is embedded with Six Sigma deployment as Lean Sigma and implemented as a company wide holistic process. There has also been partial success in a stand alone process or department. If an organisation is fighting for survival a focused but segmented approach for quick results is justified, but the recommended approach for sustainable results must be across the total value chain. With the renewed attention to 'green supply chain' (see Basu and Wright (2007), Lean and Green are here to stay and their application should be carefully thought through, otherwise partial efforts may reduce to 'flavour of the month'.

Summary of Part 3

Part 3 covers the techniques of problem solving in which generally a combination of tools are used.

The quantitative techniques in Chapter 9 are primarily aimed at advanced applications by experts (e.g. 'Black Belts' in Six Sigma programmes). It is emphasised, by taking a Pareto analogy, that 80% of the problems can be addressed without the advanced techniques described in this chapter, with the possible exception of DMAIC (Define, Measure, Analyse, Improve, Control). We have included DMAIC in this chapter in order to make it a grouping for a Six Sigma programme. The advanced techniques are more effective to improve the remaining and challenging 20% of the problems. The special emphasis on training is a pre-requisite for their successful applications.

The qualitative techniques in Chapter 10 are applicable to most operations and organisations regardless of their size or whether they are in manufacturing or service, or in private or public sector. There are training requirements for qualitative techniques but the focus must be on group working.

Part 3: Questions and exercises

1. Describe how the Failure Mode and Effects Analysis (FMEA) technique can be applied in:
   a. Improving the Customer Satisfaction of a passenger airline
   b. Risk management in the manufacture of mobile phone handsets
   c. Launching a Six Sigma project
2. In the context of Statistical Process Control (SPC), illustrate with examples the distinction between
   a. Special (or assignable) and common (or unassignable) causes of variation
   b. Variable and Attribute data
   c. UCL/LCL and USL/LSL

   What are the benefits and pitfalls of the SPC technique?

   Can you apply SPC in service industries?
3. Discuss the application of the Quality Function Deployment (QFD) technique in the following areas:
   a. Development of impulse ice-cream products
   b. Determining performance measures related to customer needs in a distribution company
   c. Customer Service in the ATM operation (cash points) of a high street bank network
4. Explain with an illustrative example how you would analyse the relationships in Design of Experiments (DOE) between key performance measures and input variables. Describe the differences between the
   - Trial and error method
   - Full factorial method
   - Fractional factorial method

   Suggest the areas of application for the above methods of DOE.
5. Compare and contrast the methodologies of Design for Six Sigma (DFSS) and DMAIC. Discuss the following statements:
   a. DFSS is appropriate for new products and processes while DMAIC is effective for improving existing processes.
   b. DFSS is resource intensive and cannot be justified for small and medium enterprises (SMEs).
   c. DFSS should be considered after accomplishing DMAIC.
6. A large retail chain operates in Manchester, Glasgow, London (three stores), Paris and Milan. As an internal consultant, recommend how you would carry out a benchmarking study and use the results in driving a performance improvement programme.
7. The Manufacturing and Supply Division of a Global Fast-Moving Consumer Group (FMCG) Company with 52 sites in 36 countries and four major product categories wishes to implement a Balanced Scorecard (BSC) based upon a performance management system.
   a. – Describe up to 10 Key Performance Indicators (KPIs) and the key steps for their implementation.

8. Compare the key features of European Foundation of Quality Management (EFQM), (International Standard Organization) ISO 9000:2000 and Malcolm Baldridge National Quality Award (MBNQA) models. Explain with reasons which model you would recommend for:
   a. A large financial services organisation in Germany
   b. A multi-national pharmaceutical company with headquarters in New Jersey
   c. A medium-sized distribution company in the United Kingdom.
9. The criteria of an EFQM Excellence Model enables an organisation to carry out self-assessment reviews. Explain how you would:
   a. Sell the idea to Senior Management
   b. Develop the self-assessment checklist and methodology
   c. Implement and maintain the process
10. A manufacturer of disc brake components has a full order book and adequate manufacturing capacity. However its customer service is unsatisfactory as shown by the average order fill efficiency for the current year at only 82%.

Explain how the Sales and Operations Planning (S&OP) technique can set up a process to improve the order fill efficiency.

Part 4
Implementation

Introduction to Chapters 11 and 12

The final part of the book brings together the common themes of previous chapters and demonstrates how the tools and techniques are implemented in practical conditions to improve quality and business performance. The last thing we want to do is to complicate the life of a manager who is actually in the arena. The description and application of individual tools and techniques are interesting but how does it help my business and how do they work together? We have addressed this question in the section in the following two chapters:

Chapter 11: Making it happen
Chapter 12: Case studies

In Chapter 11 we have described the steps and methodology of implementing a quality programme with appropriate tools and technique depending on the type, size and status of the organisation.

Chapter 12 includes a few practical case examples from a wide spectrum of businesses so that a business manager can gain a practical insight of implementation.

11

Making it happen

Quality is not something you install like a new carpet or a set of book shelves. You implant it. Quality is something you work at. It is learning process.

— W. Edward Deming

Introduction

When we began to write the chapter on implementation in 'Quality Beyond Six Sigma' (Basu and Wright, 2003) I had just returned from a Six Sigma conference at the Café Royal London. A women delegate at the conference commented 'A Six Sigma programme is like having a baby, very east to conceive but difficult to deliver'. The implementation of Six Sigma, FIT SIGMA™ and for that matter the implementation of any change programme is like 'having a baby'. It is very pleasant to conceive, but the delivery of change is difficult. According to Carnall (1999), 'the route to such changes lies in the behaviour: put some people in new settings within which they have to behave differently and, if properly trained, supported and rewarded their behaviour will change. If successful this will lead to mindset change and ultimately will impact on the culture of the organisation'.

Selection of tools and techniques

Regardless of the quality programme that an organisation may choose to adopt, a selection of tools and techniques will be essential to progress the initiative. Now the big question is which tools and techniques we need to start with? Here are three simple tips.

First, you should have a complete toolbox at hand. Just as a good golfer will not compete in a championship with just an iron, we recommend that you should acquire a full bag of irons (tools) and clubs (techniques). That is you should know how to get access to the full set. The tools and techniques are not new, they should be brought together to provide a well-stocked toolbox appropriate for the purpose. Chapters 4–10 will assist you to do just that.

Secondly, we should acknowledge that owning the best of golf clubs does not make one a Tiger Wood or by borrowing the racket of Pete Sampras does not make one to win a Wimbledon title. The success depends on the skills of the players developed by rigorous training and practice. Therefore, it goes without saying that the employees from the top to bottom of an organisation, especially the key players, should be provided with the right level of education and training to ensure that their knowledge and understanding for the tools and techniques are appropriate for the specific process and application. We have provided general guidelines of training requirements for each tool and technique in preceding chapters.

Thirdly, start with simple tools first. The experience of organisations who have successfully applied Six Sigma, such as Allied Signal, Bombardier, General Electric, Dow Chemical, to name a few, showed that 80% of their projects used 'basic tools'. The success of Total Quality Management (TQM) in Japan also indicated, according to Ishikawa (1985), that 'simple' methods led to 90–95% of process improvements. So what are these simple tools to start with? Basu and Wright (2003) suggest a set of appropriate 'FIT SIGMA™ tools'. We suggest a 'starter pack' which could be used in training workshops for 'Leadership', 'Awareness' and 'Green Belt':

IPO Diagram
Project Charter
Flow Process Chart
Process Map
Histogram
Scatter Diagram
Cause and Effect Diagram
Control Chart
Pareto Chart
Five Why
Five S
Brainstorming
Gantt Chart
Radar Chart.

Only the Control Chart from the above list would require a good understanding of statistical process control. Depending on the interest and level of participants the Run Chart could replace the Control Chart. It is evident that further the process and the programme are developed the more advanced the tools and techniques become. For example, a step change from four or five Sigma to Six Sigma or moving to Design for Six Sigma (DFSS) would certainly require the use of advanced tools and techniques. These advanced tools and techniques should be considered for 'Black Belts' and expert members of the project team. Please also see 'Introduction to Statistics' in Appendix 2.

Quality programmes

The knowledge of the tools and techniques is only one part of implementing a quality programme. Without a focused programme the training activities will not deliver results. This will only create a situation similar to a group of trained craftsmen waiting for work. Or this could be like a team of golfers waiting at the 'nineteenth hole' but there is no tournament. Whatever the name, objectives and scope of the programme could be, it is essential that you will need a programme to make the best use of your toolbox. We have adapted a FIT SIGMA™ programme from Basu and Wright (2003) to illustrate how to 'make it happen' and implement a quality programme aided by the tools and techniques.

Then implementation of FIT SIGMA™ is a major change programme designed to transform an organisation. This transformation can come about only if the cultural change of mindset is combined with facts transfer. FIT SIGMA™ is not an ad hoc localised improvement project, it is a holistic programme across the whole organisation. Therefore the essential characteristics of a FIT SIGMA™ implementation programme are:

1. Top led with totally committed management, and bottom driven.
2. Project Management discipline of Scope, Time and Budget.
3. Rigorous specialist training.
4. Company-wide open communication, spanning all functions.
5. Measurement of savings and success.

In the implementation of Lean Thinking the pitfalls of a piece-meal approach should be carefully considered. When organisations take a piece-meal approach to implementing the principles of Lean, their partial efforts often fail. If an organisation is fighting for survival a focused but segmented approach for quick results is justified, but the recommended approach for sustainable results must be across the total value chain.

In the preceding chapters we have already touched upon many factors related to implementation programmes. For example, Chapter 3 shows the four critical success factors of implementing tools and techniques and Chapters 4–8 outline the DMAIC principles of implementing Six Sigma. There are also many publications and articles relating to strategic change management and project management (e.g. Turner et al., 1996; Carnall, 1999) that would be helpful to read before embarking on a FIT SIGMA™ programme. Some other literature, however, implies that if an organisation follows a recommended systematic structured approach change management is straightforward. Our experience shows that a rigid structured approach is far from a guarantee to success. We recommend the following of a proven path with some degree of flexibility, taking into account the requirements and the existing culture of the organisation.

Below we outline proven pathways for implementing FIT SIGMA™ for organisations that are in different stages of sigma awareness and development. We have categorised three stages of development:

1. New starters of FIT SIGMA™.
2. Started Six Sigma, but stalled.
3. Completed Six Sigma, but where to now?

Implementation for new starters

At this stage the management understands the need for change and the need for an improvement programme. The main concern will be the change required to the culture of the organisation and the absence of a proven structure for transformation of a culture. Management knows what they want but how do they convince the staff that they (the staff) need to or want to change? You can take a horse to water but how do you make it drink?

Here we provide a total proven pathway for implementing a FIT SIGMA™ programme, fro the start of the initiative to the embedding of the change to a sustainable, 'bottom driven', organisation-wide culture.

Note the entry point, and the emphasis on each step of the programme could vary depending on the 'state of health' of the organisation.

The framework of a total FIT SIGMA™ programme is shown in Figure 11.1 and described below.

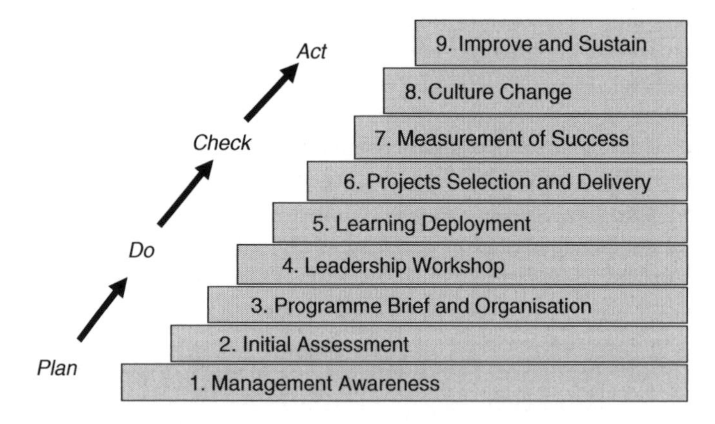

Figure 11.1 Framework of a FIT SIGMA™ Implementation (© Ron Basu).

Step One: Management Awareness

A middle manager has been tasked by the CEO with leading a Six Sigma programme in a large organisation with no previous experience of Six Sigma. The CEO has just read an article concerning Jack Welch's successes with Six Sigma in General Electric, and he is full of enthusiasm and has high expectations.

The middle manager has grim for bodings of failure. He realises that the CEO is a powerful member of the board, but after all he is only one member. In another organisation the quality manager for a medium-sized company has attended a Six Sigma conference and has mixed feelings of optimism and doubts. Our experience is that both these managers are right to be concerned.

Our experience is that it is essential to convince the CEO and at least a third of the board with the scope and benefits of FIT SIGMA™, before launching the programme. The success rate of a 'back door' approach without the endorsement of the key players cannot be guaranteed. If a programme is not company wide and wholly supported by senior management it is not FIT SIGMA™. It may be a departmental improvement project but it is not FIT SIGMA™. In cricketing terms a CEO can open the batting, but a successful opening stand needs a partner at the other end.

Our experience suggests that Management Awareness has been a key factor in successful application of Six Sigma in large organisations. Various methods have been followed including:

1. Consultants presentation to an off-site board meeting (e.g. General Electric).
2. The participation of senior managers in another organisation's leadership workshop (e.g. GSK and Ratheon).
3. Study visits of senior managers to an 'experienced' organisation (e.g. Noranda's visit to General Electric, DuPont and Alcoa).

Small and medium sized can learn from the experience of larger organisations, and indeed there can well be mutual benefits for the larger organisation through an exchange of visits. Service industry organisation could well benefit by exchanging visits with successful Six Sigma companies in the finance sector such as American Express, Lloyds TSB and Egg Plc.

During the development of the Management Awareness phase, it is useful to produce a board report or 'white' paper summarising the findings and benefits. This paper has to be well written and concise. It should not be rushed. Allow between 4 and 12 weeks for fact finding, including visits, and the writing of the 'white' paper.

Step Two: Initial Assessment

Once the agreement in principle from the board is achieved, we recommend an initial 'health check' of the organisation. There are many good reasons for carrying out an initial assessment before formalising a FIT SIGMA™ programme. These include:

1. Having a destination in mind, and knowing which road to take, is not helpful until you find out where you are.
2. Once you know the organisation's needs through analysis and measurement of the initial size and shape of the business and its problems/concerns or threats, techniques of FIT SIGMA™ can be tailored to meet the needs.

3. The initial assessment acts as a spring board through bringing together a cross-functional team and reinforces the 'buy in' at the middle management level.
4. It is likely that most organisations will have pockets of excellence along with many areas where improvement is obviously needed. The initial assessment process highlights these at an early stage.
5. The health check must take into account the overall vision/mission and strategy of the organisation, so as to link FIT SIGMA™ to the key strategy of the board. Thus the health check will serve to reinforce or redefine the key strategy of the organisation.

There are two essential requirements leading to the success of the assessment (health check) process:

1. The criteria of assessment (checklist) must be holistic covering all aspects of the business and specifically address the key objectives of the organisation.
2. The assessing team must be competent and 'trained' in the assessment process (whether they are internal or external is not a critical issue).

It is sensible that the assessment team be trained and conversant with basic fact finding methods, such as used by Industrial Engineers. Some knowledge of the European Foundation for Quality Management, EFQM, or the Baldridge (performance excellence) method of appraisal would be most useful.

Once the health check assessment is completed, a short report covering strengths and areas for improvement is required. We stress that the report should be short (not the 75 page report required for EFQM). In writing the report the company might require the assistance of a Six Sigma consultant. The typical time needed for the health check is 2–6 weeks.

Step Three: Programme Brief and Organisation

This is the organisation phase of the programme requiring a clear project brief, appointment of a project team and the development of a project plan; 'major, panic driven changes can destroy a company, poorly planned change is worse than no change' (Basu and Wright, 1998).

The programme must clearly state the purpose, scope objectives, benefits, costs, and risks associated with the programme. A FIT SIGMA™ programme is a combination of Total Quality Management, Lean Management, Six Sigma and Culture Change management. It is a big undertaking and requires the disciplined approach of project management. 'Programme management of a portfolio of projects that change organisations to achieve benefits that are of strategic benefits that are of strategic importance' (CCTA, 1999).

One risk at this stage is that management might query the budget for the programme, and there might be some reluctance to proceed. If this is the case, then it is obvious that management has not fully understood the need for change. This is why we have stressed the importance of the first step, 'Management Awareness'. Reinforcement could, however, well be needed during Step Three underpinned with informed assumptions and data including cost/benefit/risk analysis. Unless management is fully committed there is little point in proceeding.

There is no rigid model for the structure of the FIT SIGMA™ team. Basic elements of a project structure for a major change programme can be found in Basu and Wright (1998) or Turner et al. (1996). Our suggested FIT SIGMA™ model is shown in Figure 11.2.

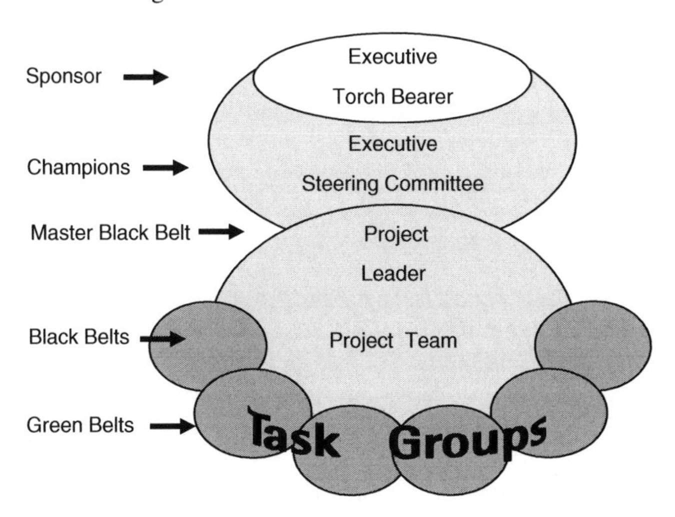

Figure 11.2 FIT SIGMA™ Programme Organisation.

Executive Torch Bearer

Figure 11.2 shows an Executive Torch Bearer. The Executive Torch Bearer ideally will be the Chief Executive (CEO) and will be the official sponsor for FIT SIGMA™. The higher up the organisation the Torch Bearer is the higher the success of the programme. The role of the Torch Bearer is to be the top management focal point for the entire programme and to chair the meetings of the Executive Steering Committee. Being a Torch Bearer may not be a time-consuming role, but it is certainly a very important to give the programme high focus, to expedite resources and to eliminate bottlenecks.

Executive Steering Committee

To ensure a high level of commitment and ownership to the project, the steering committee should be drawn from members of the board plus senior management. Their role is to provide support and resources, define the scope of the programme consistent with corporate goals, set priorities and consider and approve the programme team recommendations. In Six Sigma terminology they are the champions of processes and functional disciplines.

Programme Leader

The Programme Leader should be a person of high stature in the company – a senior manager with broad knowledge of all aspects of the business and with good communication skills. He or she is the focal point of the project and also the main communication link between the Executive Steering Committee and the programme team. Often the Programme Leader will report directly to the Torch Bearer.

The Programme Leader's role can be likened to that of a consultant. The role of the Leader is to a great extent similar to Hammer and Champy's 'czar' in *Reengineering the Corporation* (1993). The Programme Leader's role is to:

- provide necessary awareness and training for the project team, especially regarding multi-functional issues;
- facilitate work of various project groups and help them develop and design changes;
- interface across-functional departments.

In addition to the careful selection of the Programme Leader, two other factors are important in forming the team. First, the membership size should be kept within manageable limits. Second, the members should have not only analytical skills but also in-depth knowledge of the total business covering, marketing, finance, logistics, technical and human resources. The minimum number of team members should be three, and the maximum number should be seven. Any number more than this can lead to difficulties in arranging meetings, communicating and in keeping to deadlines. The dynamics within a group of more than seven people allows a pecking order to develop and for subgroups to develop. The team should function as an action group, rather than as a committee that deliberates and makes decisions. Their role is to:

- provide objective input into the areas of their expertise during the health check stage;
- lead activities when changes are made.

For the Programme Leader the stages of the project include:

- Education of all the people of the company.
- Gathering the data.
- Analysis of the data.
- Recommending changes.
- Regular reporting from the Executive Steering Committee and to the Torch Bearer.

Obviously the Programme Leaders cannot do all the work themselves. A Programme Leader has to be the type of person who knows how to make things happen and who can motivate other people to help make things happen.

Programme Team

The members of the Programme Team represent all functions across the organisation and they are the key agents for making changes. The members are carefully selected from both line management and functional background. They will undergo extensive training to achieve Black Belt standards. Our experience suggests that a good mix of practical managers and enquiring 'high flyers' will make a successful project team. They are very often the process owners of the programme. Most of the members of the Programme Team are part-time members. As a rule of thumb no less than 1% of the total workforce should form the Programme Team. In smaller organisations the percentage will of necessity be higher so that each function or key process is represented.

Task Groups

Task Groups are spin-off teams formed on an ad hoc basis to prevent the Programme Team getting bogged down in details. A Task Group is typically created to address a specific issue. The issue could be relatively major such as Balanced Score Card, or relatively minor such as investigation of losses in a particular process. By nature Task Group members are employed directly on the programme on a temporary basis. However, by providing basic information for the programme, they gain experience and Green Belt training. Their individual improved understanding and 'ownership' of the solution provide a good foundation for sustaining future changes and ongoing improvements.

Time Frame

A preliminary time plan with dates for the milestones is usually included in the Programme Brief.

We have now covered the first three Plan steps as shown in Figure 11.1.

The Do steps

In Figure 9.3 after the Plan phase there is the Do phase.

Once the programme and project plan have been agreed by the Executive Steering Team, there should be a formal launch of the programme. It is critical that all stakeholders, including managers, employees, unions, key suppliers and important customers are clearly identified. A high profile launch targeted at stakeholders such as these is desirable.

Organisation for small and medium enterprises

The organisation structure of the programme will vary according to the nature and size of the organisation, for small and medium-sized enterprises (SMEs) a typical structure is as shown in Figure 11.3. For small enterprises the Programme Leader might be part time. In all other cases the Programme Leader will be full time.

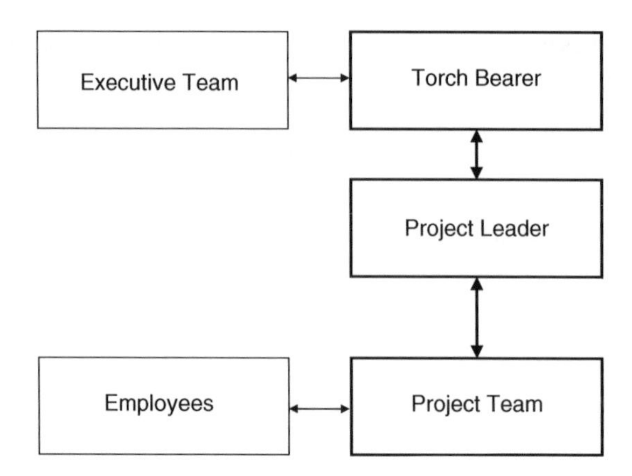

Figure 11.3 FIT SIGMA™ structure for SMEs (© Ron Basu).

Step Four: Leadership Workshop

All board members and senior managers of the company need to learn about the FIT SIGMA™ programme before they can be expected to give their full support and input into the programme. Leadership Training is a critical success factor. Step Four, Leadership Workshops, can begin simultaneously with Step One, but should be completed before Step Five, see Figure 11.1.

The Leadership Workshops will last between 2 and 5 days and will cover the following issues:

1. What is Six Sigma and FIT SIGMA™.
2. Why we need FIT SIGMA™.
3. What will it cost and what resources will be required.
4. What will it save, and what other benefits will accrue.
5. Will it interrupt the normal business.
6. What is the role of the Programme Leader and the Executive Committee.

Step Five: Training Deployment

The training programme, especially for the team members, is rigorous. One might query is it really necessary to train to achieve Black Belt certification. Indeed formal certification might not be essential. However, there is no doubt that without in-depth training of key members of the programme little value will be added in the short term, and certainly not in the long term. The training/learning deployment creates a team of experts. It as already expected that programme members will be experts in their own departments and processes as they are currently being run. It has expected that they will have the capability of appreciating how the business as a whole will be run tomorrow. FIT SIGMA™ will give them the tools for the business, as a whole, to achieve world-class performance.

It is emphasised that apart from the rigorous training in techniques and tools, the training will change the members' perceptions when they look at things. Training is an enabler not only to understand the strategy and purpose of change but as evidenced by the experience of American Express will help members to identify:

Project replication opportunities
Leveraging the results of the programme
Identification and elimination areas of rework
Drivers for customer satisfaction
Leverage of FIT SIGMA™ principles into new products and services.

Smaller organisations are very often concerned about the cost of training, especially the money paid out to consultants and for courses. In a FIT SIGMA™ programme training costs can be minimised by careful selection of specialist consultants and through the development of own training programmes. Porvair Ltd is an example of a smaller enterprise, which achieved good results with a limited training budget.

Case Example Porvair Limited

Porvair is a manufacturing company based in Wrexham, England. It currently employs 80 people and the annual sales are six million pounds. There are 2000 part numbers in the product range. The product is plastic, bronze or stainless steel, and is applied to porous media and filtration equipment. The customers are involved in high temperature catalyst recovery, medical applications, nylon spinning and water filtration.

Prior to introducing Six Sigma, the company had a respectable reputation as demonstrated by the fact that the Welsh Development Agency designated the company as a benchmark site. However the ad hoc improvement programme relied heavily on one person, the technical director. The other concern was that Porvair was experiencing poor delivery performance – less than 50% of deliveries were on time. Following a Six Sigma programme that commenced in May 2000, by March 2002 the company had achieved remarkable results, as given in Table 11.1.

Table 11.1 Porvair performance increase

	May 2000	March 2002
Delivery on time	Less than 50%	90–95%
Head count	135	80
Customer complaints (per month)	12	7
Waste	14%	10%

Plus benefits from three Six Sigma projects 206 000 pounds

The company deployed a specialist Six Sigma consultant from Belfast and the cost of training course was:

One champion	14 000 pounds
One black belt	17 000
Four Green belts	14 000
Other Costs	12 000
Total	57 000

The Black Belt (once trained) carried out further training in-house for additional Green Belts and awareness training for all employees. The training cost at Porvair equates to 1000 pounds per employee per annum.

Step Six: Project Selection and Delivery

The Project Selection process usually begins during the Training Deployment Step.

Project selection, and subsequent delivery, is the visible aspect of the programme. A popular practice is to begin by having easy, and well publicised, successes (known as harvesting hanging fruit). We recommend that quick 'wins' should be aimed for ('just do it') projects.

In a similar fashion Ericsson AB applied a simplified 'Business Impact' model for larger projects. They categorise projects under three headings:

1. Cost takeout
2. Productivity
3. Cost avoidance.

A variable weighting is allocated to each category as shown in Table 11.2.

When sufficient data is not available to provide an accurate estimate of Business Impact, an approach of 'Derived Importance' based upon scores for various categories is a practical alternative, as shown in Table 11.3.

For smaller, 'just do it' projects it is good practice to establish an 'Ideas Factory' to encourage Task Groups and all employees to contribute to savings and improvement. Very often small projects from the 'Ideas Factory' require negligible funding.

Project Review and Feedback

One important point of the Project Selection and Delivery Step is to monitor the progress of each project and to control the effects of the changes so that expected benefits are achieved. The Programme Leader should maintain a progress register supported by a Gantt Chart, defining the change, expected benefits, resources, time scale and expenditure (to budget), and show people responsible for actions.

Table 11.2 Categories of savings

Level	Cost take out	Productivity and growth	Cost avoidance
Definition	'Hard' savings – Recurring expense prior to Six Sigma – Direct costs	'Soft' savings – Increase in process capacity so you can 'do more with less', 'do the same with less', 'do more with the same'	Avoidance of anticipated cost or investment which is not in today's budget
Example	– Less people to perform activity – Less $ required for same item	– Less time required for an activity – Improved machine efficiency	– Avoiding purchase of additional equipment – Avoiding hiring contractors
Impact	Whole unit	Partial unit	Not in today's cost
Weighting	100%	50%	20%

Business impact = cost takeout + 0.5 productivity + 0.2 cost avoidance − implementation cost

Table 11.3 Derived importance of projects

Projects	Cost takeout High 10 Low 1	Productivity High 5 Low 1	Cost avoidance High 2 Low 1	Employee satisfaction High 3 Low 1	Current performance High 1 Low 10	Feasibility High 10 Low 1	Delivered importance Maximum score 40
Design packaging	9	2	0	2	7	8	20
Improve OEE	5	4	2	2	10	10	33
Passes control	1	2	2	1	5	9	20
New product development	7	1	2	3	2	2	17

This phase of review and feedback involves a continuous need to sustain what has been achieved and to identify further opportunities for improvement. It is good practice to set fixed dates for review meetings of:

Milestone Review (At least Quarterly)
= Executive Steering Committee
Torch Bearer
Programme Leader
Programme Review (Monthly)
= Programme Leader and Team,
with a short report to Torch Bearer

The problems/hold-ups experienced during projects are identified during the Programme Review with the aim of the Project Team taking action to resolve sticking points, and if necessary requests are made of the Executive Steering Committee for additional resources.

Step Seven: Measurement of Success

The fundamental characteristic of a Six Sigma or FIT SIGMA™ programme that differentiates it from a traditional quality programme is that it is results orientated. Effective measurement is the key to understanding the operation of the process and it forms the basis of all analysis and improvement work. In a construction project the milestones are tangible, they are physically obvious, but in a change programme such as FIT SIGMA™ the changes are not always apparent. It is important to measure, display and celebrate the achievement of milestones in a FIT SIGMA™ programme. We emphasise the importance of performance management to improve and sustain the results of a FIT SIGMA™ programme. The process and culture of measurement must start during the implementation of changes.

Our experience is that the components of measurement of success should include:

Project tracking
FIT SIGMA™ metrics
Balanced Score Card
Self-assessment review (Baldridge or EFQM).

There are software tools available such as Minitab (www.minitab.com) and KISS for detailed tracking of larger Six Sigma projects. However in most programmes the progress of savings generated by each project can be monitored on an Excel spreadsheet. We recommend that summaries of results are reported and displayed each month. Examples of forms of displays are shown in Figures 11.4 and 11.5.

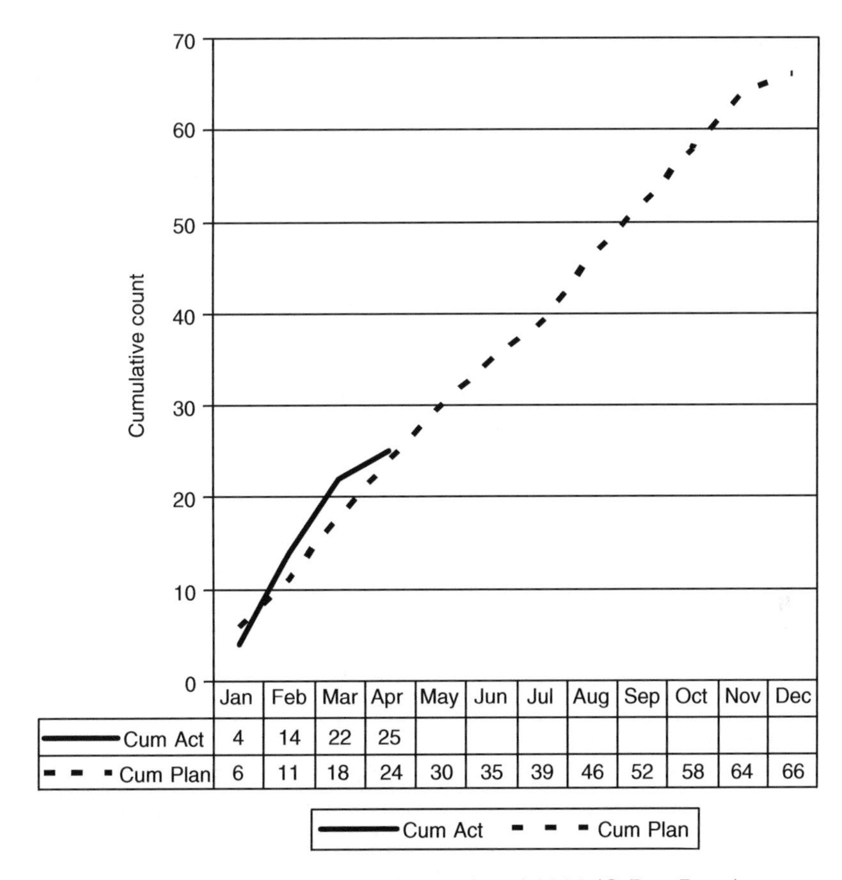

	Jan	Feb	Mar	Apr	May	Jun	Jul	Aug	Sep	Oct	Nov	Dec
Cum Act	4	14	22	25								
Cum Plan	6	11	18	24	30	35	39	46	52	58	64	66

Figure 11.4 Projects planned and completed 2002 (© Ron Basu).

FIT SIGMA™ metrics

FIT SIGMA™ metrics are required to analyse the reduction in process variance and the reduction in the rate of defects resulting from the appropriate tools and methodology.

A word of caution. Black Belts can get caught up with the elegance of statistical methods and develop a statistical cult. Extensive use of variance analysis is not recommended.

The following FIT SIGMA™ metrics are useful, easy to understand and easy to apply:

Cost of Poor Quality = COPQ ratio
Defects per Million Opportunities = DPMO
First Pass Yield = FPY.

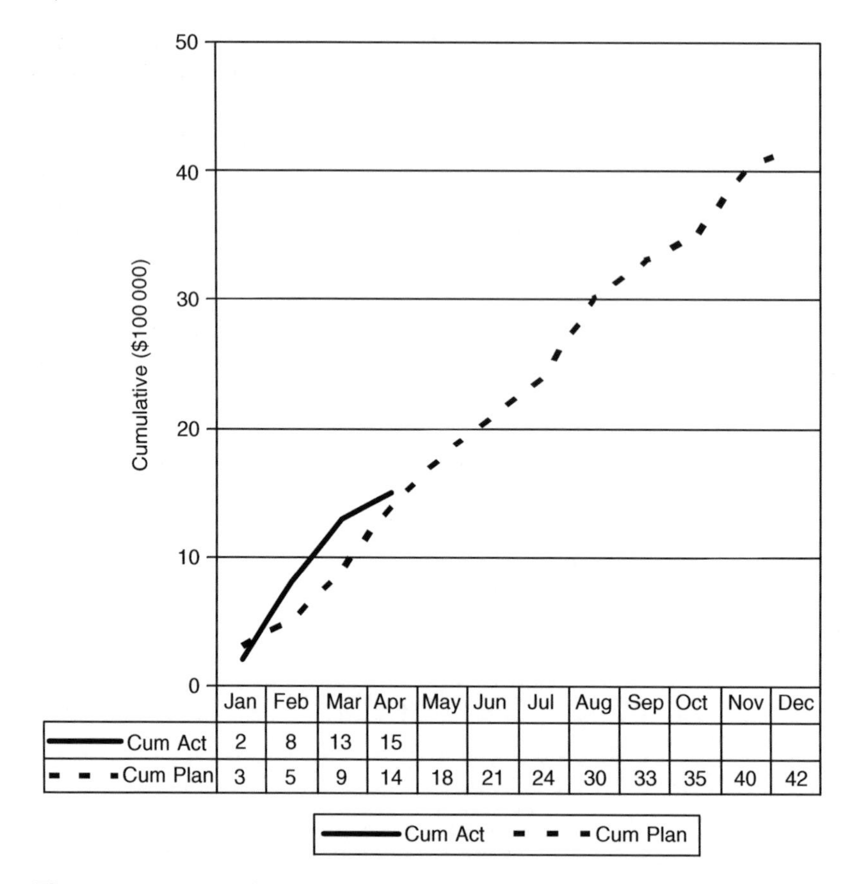

	Jan	Feb	Mar	Apr	May	Jun	Jul	Aug	Sep	Oct	Nov	Dec
Cum Act	2	8	13	15								
Cum Plan	3	5	9	14	18	21	24	30	33	35	40	42

Figure 11.5 Value of planned and completed projects (© Ron Basu).

COPQ

$$\frac{\text{Internal Failure \$} + \text{External Failure \$} + \text{Appraisal and Prevention \$} + \text{Lost Opportunity \$}}{\text{Monthly Sales \$}}$$

DPMO

$$\frac{\text{Total number of defects} \times 1\ 000\ 000}{\text{Total units and opportunities per unit}}$$

FPY

$$\frac{\text{Number of Units Completed Without Defects and Rework}}{\text{Number of Units Started}}$$

From measuring and monitoring FIT SIGMA™ metrics each month, opportunities for further improvement will be identified.

As we have already emphasised, a carefully designed Balanced Score Card, see Chapter 6, is essential for improving and sustaining business performance. It is generally viewed that the Balanced Score Card is applicable for a stable process and thus should be appropriate after the completion of the FIT SIGMA™ programme. This maybe so but unless the measures of the Balanced Score Card are properly defined and designed for the purpose at an early stage, its effectiveness will be limited. Therefore, we strongly recommend that during the FIT SIGMA™ programme the basics of the Balanced Score Card to manage the company-wide performance system should be established.

The fourth component of measurement is the 'Self-assessment and Review' process. There are two options to monitor the progress of the business resulting from the FIT SIGMA™ programme. They are to use:

simple checklist to assess the overall progress of the programme;
proven self-assessment process such as the European Foundation of Quality Management (EFQM) or the American Malcolm Baldridge system.

We recommend the second option. In the initial health check appraisal stage we recommended using EFQM or Baldridge, thus the methodology will have already been applied. Additionally it gives further experience in the self-assessment process, which will enable future sustainability. And finally it will provide the foundation should the organisation wish at a later stage to apply for an EFQM or Baldridge award.

Step Eight: Culture Change

A culture change must NOT begin by replacing middle management by imported 'Black Belts'. Winning over, not losing, middle management is essential to the success of FIT SIGMA™, or for that matter any quality initiative.

What is required is that the all-important middle management, and everyone else in the organisation, understand what FIT SIGMA™ is, and hence the culture of quality. An understanding of Deming's 14 points, see Chapter 2, would be a sound start.

The FIT SIGMA™ culture is shown in Table 11.4.

Table 11.4 FIT SIGMA™ culture

1. Total vision and commitment of top management throughout the programme
2. Emphasis on measured results and the rigour of project management
3. Focus on training with short-term projects and results, and long-term people development
4. Use of simple and practical tools
5. Total approach across the whole organisation (holistic)
6. Leverage results by sharing best practice with business partners (suppliers and customers)
7. Sustain improvement by knowledge management, regular self-assessment and senior management reviews

Air Academy (www.airacad.com) (April, 2002) claim that it is important to understand the culture type of an organisation to ensure the culture change necessary for the success of a Six Sigma programme. The culture types identify the following:

1. *Clan culture*: The organisation is considered to be people orientated. It is a nice place to work where people share similar interests, much like a country club, i.e. clan like, they all have similar beliefs and values.
2. *Hierarchical culture*: The organisation has a formalised top-down structure, and people are governed by rules and procedures.
3. *Enterprise culture*: The organisation is goal orientated. Results are measured and members are competitive. People are primarily concerned with getting the job done.
4. *Adhocracy culture*: The organisation is dynamic, creative and entrepreneurial. People are proactive and take risks, are innovative and look for alternatives.

FIT SIGMA™ requires a balanced culture comprising key characteristics of the above four types. If an organisation is predominantly one type, some cultural change will be required. Training Deployment, see Step Five of Figure 9.3, includes training for culture change. But training alone will not transform the mindset required for FIT SIGMA™.

Creating a Receptive Culture
An often asked question is how to change culture
It all begins with Vision.

The vision of quality must begin with the chief executive. If the chief executive has a passion for quality and continuous improvement, and if this passion can be transmitted down through the organisation, then paradoxically, the ongoing driving force will be from the bottom up rather than enforced from above, and with everyone sharing the same vision. For similar viewpoints regarding TQM see Crosby (1979), Ishikawa (1985), Schonberger (1986), Albrecht (1988), Collins and Porras (1991), Creech (1994), Dulewicz et al. (1995) and Gabor (2000).

The word 'vision' suggests almost a mystical occurrence (Joan of Arc) or an ideal (such as expressed by Martin Luther King 'I have a dream'). The same connotation is found when looking at vision in the organisational context. A leader with a vision is a leader with a passion for an ideal. 'But, unless the vision can happen, it will be nothing more than a dream' (Wright, 1996, p. 20, also see El-Namki, 1992; Langeler, 1992). To make a vision happen within an organisation, there has to be a cultural fit. Corporate culture is the amalgam of existing beliefs, norms and values of the individuals who make up the organisation ('the way we do things around here') (Peters and Waterman, 1982; Peters and Austin, 1986). The leader may be the one who articulates the vision and makes it legitimate, but unless it mirrors the goals and aspirations of the members of the organisation at all levels, the vision won't happen (Albrecht, 1988). As Stacey (1993, p. 234) says 'the ultimate test of a vision is if it happens'.

Culture and values are deep seated and may not always be obvious to members. As well as the seemingly normal aversion to change by individuals, often there is a vested interest for members of an organisation to resist change. Middle management often is more likely to resist change than are other members. Machiavelli (1513) wrote 'It must be considered that there is nothing more difficult to carry out, nor more doubtful to success, nor more dangerous to handle, than to initiate a new order of things'. Human nature hasn't really changed much since the 16th century!

Organisations are made up of many individuals, each with their own set of values. The culture of an organisation is how people react or do things when confronted with the need to make a decision. If the organisation has a strong culture, each individual will instinctively know how things are done and what is expected. Conversely, if the corporate culture is weak, the individual may not react in the manner that management would hope (Peters and Waterman, 1982; Peters and Austin, 1986; Carnall, 1999).

To engineer or change a culture there has to be leadership from the top. Leading by example might seem to be a cliché but, unless the chief executive can clearly communicate and demonstrate a clear policy by example, how will the rest of the people know what is expected? (also see Foreman and Money, 1995; Foreman, 1996, 2000). Leadership does not have to be charismatic, but it has to be honest. Leadership does not rely on power and control. Basu and Wright (1998) find that real leaders communicate face to face, not by memos.

Mission statement to signal change

A new mission statement would seem to be a logical way for a chief executive to signal a change in direction for an organisation.

Ideally the mission statement should be a true statement as to the reason for being of the organisation. It should be realistic and state the obvious. Profit is not a dirty word (Friedman, 1970) and customer service is important (Zeithaml et al., 1990; Kotler, 1999). Generally the key resource lacking in many organisation is quality people (Barlett and Ghoshal, 1994; Mintzberg, 1996; Knuckey et al., 1999). Therefore it would seem obvious for any mission to say we are in business to make a profit, and we will make a profit by providing the customers with what they want, and we recognise that our most important resource in making this mission happen is our people. It is important that the new mission be in tune with what the people of the organisation believe (the culture). To achieve a mission that fits the culture it would seem sensible to get the involvement and interest of all the staff in writing the new mission. Thus in this manner a change in culture could begin with the determination and the buy- in by the staff into the new mission.

Learning for Change

'If employees, organisation wide, are going to accept change, and themselves individually change, they will need to learn certain skills such as:

Understanding work processes
Solving problems

Making decisions
Working with others in a positive way.

All these types of skills can be taught. The main message that has to be learnt is the need for cultural change, and for people to trust each other. In particular management has to win the trust of lower level staff and has to learn how to change from autocratic management to coaching and mentoring. Lower level staff, in turn, have to learn to trust management' (Wright, 1999, p. 219, also see Hall, 1999; Axelrod, 2001) who express similar views. Once this has been achieved the culture will be such that the organisation will be in tune with the philosophy of FIT SIGMA™, and the Ninth Step of FIT SIGMA™, Improve and Sustain, will be second nature.

Communication

Finally, the key to sustaining a FIT SIGMA™ culture is communication. Methods of communication include:

- Intranet FIT SIGMA™ website, specifically developed, or clearly visible in the corporate website.
- Specially produced videos or CD Discs.
- FIT SIGMA,™ monthly news letter.
- Internal e-mails, voicemails, memos with updated key messages – NOT slogans such as work smarter not harder and other tired clichés.
- Milestone celebrations.
- Staff get-togethers, such as special morning teas, Friday afternoon social hour, 'town hall' type meetings.
- 'Ideas factory' or 'think tank' to encourage suggestions and involvement from employees.

Step Nine: Improve and Sustain

Improve and sustain is the cornerstone of a FIT SIGMA™ programme. This is similar to the fifth stage of team dynamics for project teams (Forming, Storming, Norming, Performing and Mourning). In the Mourning stage the project team disbands and members move onto other projects or activities. They typically regret at the end of the project and break up of the team, and the effectiveness or maintenance of the new method and results gradually diminish. In Chapter 6 we have discussed in some detail to achieve sustainability and four key processes must be in place:

1. Performance management
2. Senior management review
3. Self-assessment and certification
4. Knowledge management.

The 'end game' scenario should be carefully developed long before the end of the programme. There may not be a sharp cut-off point like a project handover, and the success of the scenario is in the making of a smooth transition without disruption to the ongoing operation of the business.

As part of the performance management the improvement targets should be gradually, and continuously, stretched and more advanced tools considered for introduction. For example, the DFSS is resource hungry, and can be considered at a later stage in a FIT SIGMA™ programme. With Six Sigma the aim is to satisfy customers with robust 'zero defect' manufactured products and to do so DFSS is fully deployed covering all elements of: Manufacturing, Design, Marketing, Finance, Human Resource, Suppliers and Key Customers (including supplier's suppliers and customer's customers).

At an advanced stage of the programme milestone, review should be included in senior management operational review team meetings (such as the sales review meetings and operational planning meetings/committees), i.e. not only the FIT SIGMA™ Executive Steering Committee.

We recommend a pure play EFQM (or other form of self-assessment) should be incorporated as a 6 monthly feature of the FIT SIGMA™ programme. Even if the company gains an EFQM or Baldridge award, the process must continue indefinitely.

Two specific features of knowledge management need to be emphasised. First, it is essential that the company seeks leverage from FIT SIGMA™ results by rolling it out to other business units and main suppliers. Secondly, it is equally important to ensure that career development and reward schemes are firmly in place to retain the highly trained and motivated Black Belts. The success of sustainability of FIT SIGMA™ is when the culture becomes simply 'this is the way we do things'.

Time scale

The time scale of FIT SIGMA™ implementation will last several months and is, of course, variable. The time not only depends on the nature or size of the organisation, but also on the business environment and the resources available. Four factors can favourably affect the time scale:

1. Full commitment of top management and the board
2. Sound financial position
3. Correct culture (workforce receptive to change)
4. A competitive niche in the marketplace.

It is good practice to prepare a Gantt Chart containing the key stages of the programme and to monitor the progress. Figure 11.6 shows a typical timetable for a FIT SIGMA™ programme in a single site medium-sized company. The diagram shows an order of magnitude only, and the sequence could well vary. The timeline is not linear, stages overlap, and frequent looking back to learn for future progress should occur.

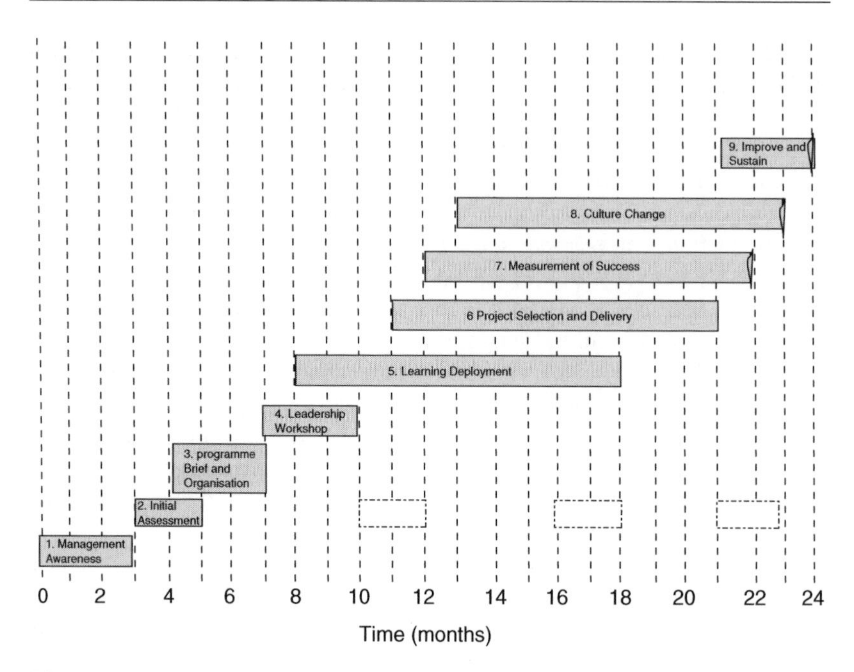

Figure 11.6 A typical time plan for a FIT SIGMA™ programme (© Ron Basu).

FIT SIGMA™ for Stalled Six Sigma

Some organisations have already attempted to implement a Six Sigma (or a TQM) programme, but it has stalled. Results are not being achieved and enthusiasm is waning; in some cases the programme has effectively been abandoned. The reasons for stalling are various but often are: an economic downturn (such as one experienced in the Telecommunications industry in 2001), a change in top management, or a merger or take over.

FIT SIGMA™, Not a Quick Fix

During restructuring, or if the company is in survival mode, the implementation of FIT SIGMA™ is not appropriate. FIT SIGMA™ is not a quick fix. After the short-term cost saving measures of a survival strategy, when the business has stabilised, and a new management team is in place, then Six Sigma can be restarted, but this time it has to be done correctly, using the FIT SIGMA™ approach.

It is likely that when restarting many of the steps including Training/Learning, deployment will not need to be repeated. However in a restart there is one big issue that makes life more difficult, and that is credibility. How do you convince all the people that it will work the second time around. This will put special pressure on Step Eight, Culture Change. The employees could well be tiered of excessive statistics and complex Six Sigma tools. Selection of appropriate tools is a strong feature of FIT SIGMA™.

The FIT SIGMA™ programme for a restarter will naturally vary according to the condition of the organisation, but the programme can be adapted within the framework shown in Figure 11.1. The guidelines for each step are:

1. Management Awareness, if not cannot restart.
2. Initial Assessment. This has to be the restart point. Where are we, where do we want to go?
3. Programme Brief and Organisation. The programme will need to be rescoped and new teams formed.
4. Leadership Workshop. This will be essential, even if management has not changed.
5. Training/Learning Deployment. Appropriate tools should be selected. If past team members, on particular Black Belts, are not happy with terminology of the old programme, new terms should be used. The title Black Belt itself is not sacrosanct and might be changed. If the 'old' experts are still in the organisation, then training time might be reduced, i.e. only 1-week workshop might be sufficient.
6. Project Selection and Delivery. Same as the full FIT SIGMA™ programme, harvest the hanging fruit.
7. Measurement of Success. Review the old measures, what worked, what didn't, and follow the full FIT SIGMA™ programme.
8. Culture Change. This will be critical. Top management support will have to be extremely obvious. Reward and appraisal systems will have to be aligned to FIT SIGMA™.
9. Improve and Sustain. Same as for FIT SIGMA™.

FIT SIGMA™ for Successful Companies

William Stavropoulos, the CEO of Dow Chemical, is reported to have once said 'the most difficult thing to do is to change a successful company'. It is true that employees of companies enjoying a high profit margin with some dominance in the market are likely to be complacent and to feel comfortable with the status quo. Perhaps it is even more difficult to stay at the top or sustain success if the strategy and processes are not adaptable to change. Darwin famously said, 'It is not the strongest species that survive, nor the most intelligent, but the ones most responsive to change'. It is possible for management of some companies, after the completion of a highly successful Six Sigma programme, find their attention diverted to another major initiative such as e-Business or Business to Business Alliances. Certainly new initiatives must be pursued, but at the same time the long term benefits that could be achieved from Six Sigma should not be lost. FIT SIGMA™ for sustainability, staying healthy, is the answer.

If a company has succeeded with Six Sigma, then this is the time to move right onto FIT SIGMA™ to achieve Step Nine, Improve and Sustain.

External Consultants

Many companies, especially SMEs, are often concerned with the cost of consultants for a Six Sigma programme. Large consulting firms and academies for Six Sigma could well expect high front end fees. With FIT SIGMA™ our approach is to be selective in the use of outside consultants. That is use outside consultants to train your own experts, and to supplement own expertise and resource when necessary. No consultant will know your own company as well as your own people do. For this reason we do not favour the use of external consultants as a Programme Leader.

In a FIT SIGMA™ programme the best use of consultants is in:

Step Two: Initial Assessment – Here we would use a Six Sigma expert or an EFQM consultant to train and guide your team.

Steps Four and Five: Leadership Workshop and Training/Learning Deployment – Outside consultants will be needed to facilitate the Leadership Workshops, and to train your own 'Black Belts'. Once trained your own 'Black Belts' will train 'Green Belts' and develop new 'Black Belts'.

Step Eight: Culture Change – An outside consultant is best suited to develop a change management plan for change of culture.

12

Case studies

The bitterness of low quality is not forgotten
Nor can it be sweetened with low price

— Marquis De Lavant (1734)

Introduction

The first wave of Six Sigma, following the grand groundwork of Motorola, included Allied Signal, Texas Instruments, Ratheon and Polaroid (to name a few). GE entered the arena in the mid-nineties and in turn was followed by many powerful corporations including SONY, DuPont, Dow, Bombadier, Ford and GSK. The success of GE Capital attracted many companies in the service sector, such as American Express, Lloyds TSB, Egg.com and Vodafone, to apply Six Sigma as a major business initiative. Some established companies like Unilever, ICI and BT have also applied Six Sigma tools and techniques but attempted to align them to existing strategic and continuous improvement programmes.

The ability to leverage the experience of successful Six Sigma players proved highly attractive both as a competitive issue and also to improve profit margins. The following case examples provide insights for organisations that are achieving success in their business performance through the use of Six Sigma or related operational excellence programmes. The objective of including these case examples is to offer a set of good proven practices through which a student or a practitioner can obtain a better insight of the application of tools and techniques and how the organisations have adopted the approach.

The case examples are taken from the real life experiences of many organisations around the world. We have not made any effort to make them tidy to suit any subject area. It is hoped that these cases provide a useful resource to stimulate discussions and training of tools and techniques in quality management.

Case Study 1: Product innovation at Elida Faberge, UK

Background

Elida Faberge, UK, is a subsidiary of Unilever Group and manufactures and supplies Personal Care products like shampoo, deodorants and tooth pastes. The company examined the way in which innovation takes place in the business and how it is managed and controlled. The purpose of this review was to develop a clearly mapped transparent process for new products delivery ahead of competition with the optimum use of resources. The relaunch of a major shampoo brand (Timotei) in March 1992 was the pilot project using this process.

The relaunch involved four variants of shampoo in four sizes and two variants of conditioner in three sizes and incorporated the main technical changes of a new dispensing cap and changes in formulations.

Approach

A multi-disciplinary task force was set up to develop the new process in readiness for a company-wide launch in April 1993. The innovation process used by Elida Faberge was based on the principles of 'phases and gates' (Figure 12.1) and consisted of:

- a logical sequences of phases and exit gates;
- the use of empowered multi-disciplinary core teams;
- using gates as key points where 'gatekeepers' review progress and decide whether a project may proceed or be terminated;
- monitoring progress at each gate and for the overall project using key performance measures.

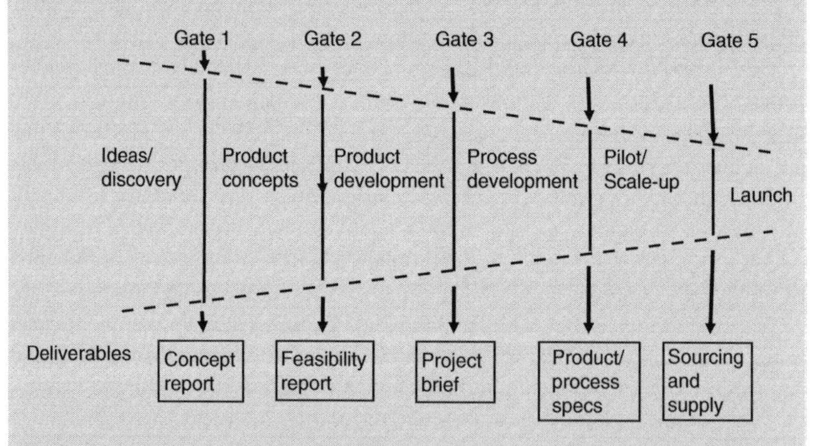

Figure 12.1 Product development cycle.

The project team had a membership representing the key processes, such as Brand Development, Supply Chain, Quality Management and Customer Development. The members received additional training in project management and the use of SPC tools. A Project Leader coordinated the team's efforts and reported the progress to the 'gatekeepers'.

Implementation

Using the multi-functional team approach as described, the Timotei project was launched and managed to the implementation stage (Phase 4) by a team lead by a Brand Manager. A second team was set up under the technical leadership of a Development Manager to see the project through to launch, in particular to set up supply lines, production trials and production start-ups.

Two other teams were also involved – a separate packaging supply chain team and a TQ team on the Timotei start-up production line. Technical audits were also set up at three packaging supply sites and corrective actions were issued. The sites were revisited to ensure that all corrective actions have been satisfactorily achieved.

Results and learning points

The Timotei project was a success both in terms of sales and as a managed innovation process. Notable problems with materials supply were not encountered. However, the considerable sales success of the product overstretched the bottle supplier, so that a second supplier was brought on stream. The quality and performance ratings of the production line achieved targets.

The key learning points of the project included:

- The use of multi-functional team allowed actions to be taken swiftly with shared ownership and resources at the team's disposal.
- The additional training of team members in project management and SPC tools ensured that the project was managed in a systematic, organised and disciplined manner.
- The effective use of phases (with multiple project leadership) and gates contributed to the development of the 'Innovation Funnel' of Unilever which was introduced to all product groups.

Tools and techniques used

Process Mapping, Cause and Effect Diagram, Control Charts, QFD and FMEA.

Case Study 2: Lean Manufacturing at aluminium industry, USA

Background

Author and researcher Dr. James P. Womack, President of the Lean Enterprise Institute, told attendees at a meeting of the Aluminium Association that aluminium producers are among the many basic industries that could do a lot more with less difficulty and lower costs if they did some lean thinking.

He described the labyrinth and costly processes associated with delivering an aircraft, automobile or drinks can, and suggested that manufacturers think about the consumer first and work backwards to gain efficiencies and to cut out the non-value adding activities.

Approach

'Ask, from the customer's standpoint, what is of value among your activities? What is wasted? How can we eliminate the waste? It's so simple and yet so very hard to do. Most people are in love with their assets, technologies and organization', Womack said. Womack cited a military aircraft programme as an example. 'The typical subassembly goes through 4 plants, 4 states and 74 organizational handoffs between engineering, purchasing and fabricating operations. It goes 7600 miles and takes darn near forever. It takes 2–4 years from beginning to flyaway condition'. Manufacturers need to identify the value stream for each of their products, and document all the steps it takes to go from raw material to the customer.

Implementation

'Step 1, get the value right', said Womack. 'Your customer is not interested in your assets. He is interested in his value. Step 2, identify the value stream from start to finish, not just within the walls of your plant or company'.

Organize the remaining value-creating steps so they impact the product in a continuous flow, he advised.

Results and learning points

'If, and only if, you can create flow, then you can move to a world of pull,' Womack said. 'You put the forecast in your shredder and get on with your life. And you make people what they want'. The customer says, 'I want a green one.' And you say, 'Here's a green one.' 'That's a very different world from your world of endless forecasts, always wrong, and the desperate desire to keep running, which makes you produce even more of the wrong thing because that makes the numbers look good in the short term.'

Tools and techniques used

Value Stream, Kanban, Flow Process Chart.

(Case reported in *Metal Center News*, May 1999, Vol. 39, Issue 6, p. 123.)

Case Study 3: Supply chain logistics at National Starch, USA

Background

National Starch and Chemical Company, based in Chicago, began measuring shipping performance on a corporate basis in early 1990s. The initial results indicated that, at best, the Adhesives Division was achieving 80–85% on-time delivery performance. This is a business where there is a need to ship a multitude of products in a variety of containers with varying lead times from customers. The logistics department set an objective to reduce shipping delays from 15% to 5% and at the same time reduce overall inventories.

Approach

The approach the division took was to classify their products in specific containers by A, B, C categories. The categories represented volume of product and the customer base purchasing the product. The lead time for shipping the product and the inventory held were related to the A, B and C designations, with A being shorter lead time with available inventory.

Implementation

A number of local cross-functional teams were formed comprising Sales, Manufacturing, Technical Service, Materials Management and Customer Service. The task of the teams was to work on ABC product classification and delivery service standards. Essentially every product in each container was evaluated based on the volume and number of customers purchasing it. The Chicago plant reviewed over 200 products in approximately 15 different containers for more than 700 orders per month.

The next task was to develop a tracking system to report on delays against standard lead times and compare them with the date agreed with the customer.

After agreement was reached on the ABC products and their safety stocks, the order patterns of these products were reviewed and minimum, maximum and re-order levels were established. The employees were thoroughly trained and the system was also presented at each regional sales meetings. It was important to gain support of sales stuff since they needed to 'sell' it to the customers.

To ensure that the ABC and safety stocks reflect changing market demands, a cross-functional team reviews the ABC product designations and inventory lists every month.

Results and learning points

The improvement in shipping performance was dramatic since the implementation of the ABC system as shown by the figures of shipping delays as follows:

Year 0 overall:	14.7%
Year 1 overall:	3.8%
Year 2 overall:	1.2%

There were other benefits in the supply chain operations and performance including:

- Sales team increased the amount of time to work proactively with customers.
- There has been a reduction in finished goods and raw materials inventory.
- Decreased the 'last minute' expedited freight charges.
- Improved the communication between departments especially between sales and logistics.

Tools and techniques used

Pareto Chart, Run Chart, S&OP

Case Study 4: Six Sigma in General Electric, USA

Background

General Electric (GE) has been at the top of the list of Fortune 500's most admired companies for the last 5 years, and without doubt their Six Sigma programme has played a key role in their continued success. In 2001 GE's turnover was over $125.8 billion, they employed 310 000 people worldwide and their market value was $401 billion. With earning growing at 10% per annum, GE also has the enviable record of pleasing Wall Street, and financial analysts year after year. GE's products and business categories span a wide spectrum of automotive, construction, health care, retail, transport, utilities, telecommunications and finance.

The CEO of GE, Jack Welch, is reported to have become attracted to the systematic and statistical method of Six Sigma in the mid-nineties. He was ultimately convinced of the power of Six Sigma after a presentation of Allied Signal's former CEO, Larry Bossidy, to a group of GE employees. Bossidy, a former Vice Chairman of GE, had witnessed excellent returns from Allied Signals experience with Six Sigma.

Approach

In 1995 GE retained the Six Sigma Academy, an organisation started by two early pioneers of the process, both ex-Motorola executives, Michael Harry and Richard Schroeder. It was pointed out that the gap between Three Sigma and Six Sigma was costing GE between $7 and $10 billion annually in scrap, rework, transactional errors and lost productivity. With the full and energetic support of Jack Welch, senior management became fully committed to the Six Sigma programme. 'GE Quality 2000' became the GE mantra for the 1990s and beyond. Jack Welch declared that 'Six Sigma, 'GE Quality 2000' will be the biggest, the most personally rewarding and in the end the most profitable undertaking in our history!' While financial benefits and the share price were a driving force in Six Sigma deployment, GE identified four specific reasons for implementing Six Sigma:

1. Cost reduction
2. Customer satisfaction improvement
3. Wall Street recognition
4. Corporate synergies.

Although Motorola pioneered the Six Sigma programme in the 1980s to improve manufacturing quality and to eliminate waste in production, GE broke the mould of Motorola's original process by applying the Six Sigma standards to its service-oriented businesses – GE Capital Services and GE Medical Systems. Note GE Capital Services accounts for nearly half of GE's total sales.

The structure of a typical GE Six Sigma team is shown in Figure 12.2.

Implementation

The Six Sigma programme was launched in 1995 with 200 separate projects supported by a massive training effort. In the following 2 years a further 9000 projects were successfully undertaken and the reported savings were $600 million. The training investment for the first 5 years of the programme was close to $1 billion. GE also instituted a personnel recruitment plan to augment the cadre of dedicated full-time Six Sigma staff.

The GE programme revolved around the following few key concepts, all focused on the customer and internal processes:

- *Critical to quality*: Determination and development of attributes most important to the customer.
- *Defect*: Identification of failure to meet customer wants.
- *Process capability*: What the process can deliver.
- *Variation*: What the customer sees and feels as against what the customer wants.

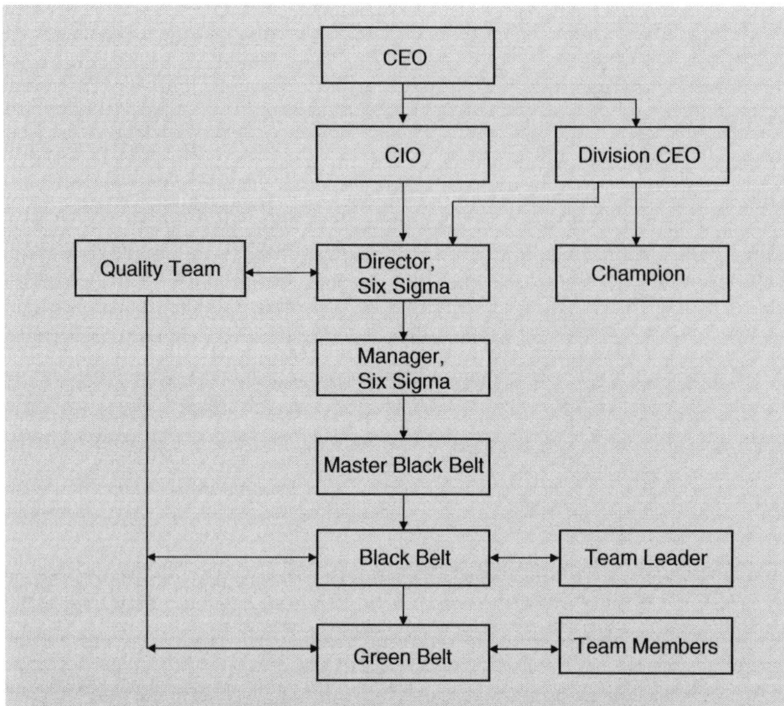

Figure 12.2 A typical Six Sigma team, GE.

- *Stable operation*: Ensuring consistent and predicable processes to improve what the customer sees and feels.
- *Design for Six Sigma*: Designing to meet customer needs and process capability.

Model for roll out

There does not appear to one universal model for rollout of Six Sigma amongst the companies that have implemented a Six Sigma programme. However, the Six Sigma Academy has advised that there is a general model which is effective and which has been adopted/developed by GE. This general model is shown in Table 12.1.

Results and learning points

Figure 12.3 shows the direct financial benefits achieved by GE over a 4-year period. This provides evidence of the very real benefits that are achievable from a Six Sigma programme. With FIT SIGMA™ the next stage is to sustain and to grow the benefits.

At the second level of benefits, where the impact on savings is not direct, the achievements, average per year, include:

- 20% margin improvement
- 12–18% capacity increase

Table 12.1 GE training model

Phase One	Business units select champions and Master Black Belts. The Six Sigma Academy recommends one champion per business group and one Master Black Belt for every 30 Black Belts
Phase Two	Champions and Master Black Belts undergo training. The overriding deployment plan is developed
Phase Three	Champions and Master Black Belts, with the assistance of Black Belts, begin identifying potential projects
Phase Four	Master Black Belts receive additional training, focusing on how to train other staff
Phase Five	Black Belts undergo training and the first projects are officially launched
Phase Six	Black Belts begin training Green Belts

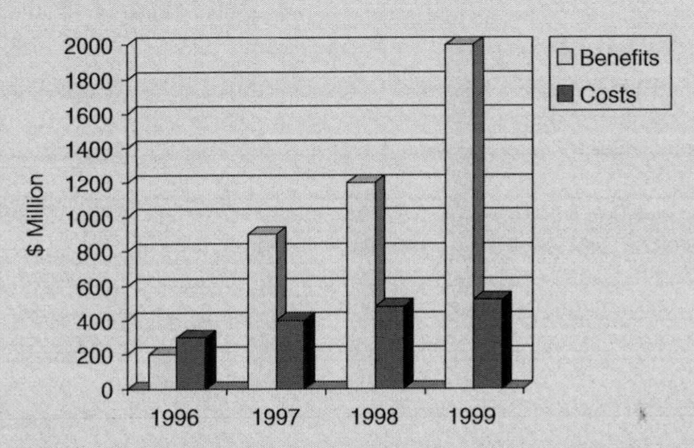

Figure 12.3 Six Sigma payoff at GE.

- 12% reduction in headcount
- 10–30% capital expenditure reduction.

Some specific examples from business units are:

- *GE Medical Systems*: In the introductory year there were 200 successful projects.
- *GE Capital*: Invested $6 million over 4 years to train just on 5% of the work force who worked full time on quality projects, and 28 000 quality projects were successfully completed.
- *GE Aircraft Engines*: The time taken to overhaul engines was reduced by an average of 65 days.
- *GE Plastics*: In just one project, a European polycarbonate unit increased capacity by 30% in 8 months.

Market consultants and analysts have reacted favourably to GE's achievements with Six Sigma. Merill Lynch is quoted as saying, 'Six Sigma balance sheet discipline plus service and global growth are helping fuel

(GE's) 13% earning per share gains'. On 8 May2002, GE announced that they will deliver record earning in 2002 of more than $16.5 billion, and was comfortably forecasting double-digit earnings growth for 2003.

At one level, to emulate GE maybe considered as being beyond the reach of many companies. It is cash rich, and its business generates over $10 billion per month ($125.8 billion sales for 2001). It makes real things like turbines and refrigerators, and people buy their products with real money. However, on a closer examination, there are some strong learning points from the GE Six Sigma programme that can benefit ANY company embarking on a quality programme. These learning points include:

1. *Leadership support*: There is absolutely no doubt from published data that the chief architect of success was Jack Welch. For any organisation wanting to change a culture such as required for Six Sigma and FIT SIGMA™, strong unstinting leadership from the top is essential. Likewise, all senior management has to be engaged and believe in the philosophy.
2. *Definition of Six Sigma objectives*: The objective is to be world class. World-class companies such as GE recognised that quality initiatives are synonymous with profit enhancement and share price. World class means internal efficiency, best practice and a focus of customer satisfaction. The lesson is that any thing less than an ambition to be world class simply won't do.
3. *Development of initial processes and tools*: At GE each problem was defined through measurement and analysis along a five-step DMAIC approach (DMAIC – Define, Measure, Analyse, Improve and Control), and the use of the seven quality 'tools', i.e. Control Charts, Defect Measurement, Pareto Analysis, Process Mapping, Root Cause Analysis, Statistical Process Control and Decision Tree Diagram. The lesson is that a structured approach has to be followed for Six Sigma process management.
4. *Alignment of Six Sigma with career paths*: At GE Black Belt status became essential for staff on the fast track for advancement. Black Belts were rewarded with share options (in most companies share options are reserved for senior management). The lesson from this is that recognition has to be given to motivate and retain valuable talent.
5. *Six Sigma and service industries*: The piloting of Six Sigma in GE Medical Systems and GE Capital Services incontrovertibly proved that Six Sigma is not just for manufacturing, the process is equally applicable to all operations including services. GE has opened the gate for service operations. In Western economies 80% of gross domestic product is from the service sector.

Tools and techniques used

Most of the tools and techniques covered in this book.
Source: Basu and Wright (2003).

Case Study 5: Integrating Lean and supply chain in Seagate, USA

Background

Seagate Technology is the world's largest manufacturer of disc drives and HDD recording media. With its headquarters at Scotts Valley, California, the company employs 62 000 people and its turnover in 2000 exceeded $7 billion. The business operates in a market environment of short product life cycle and quick ramp to high volume. The data storage market is growing 10–20% per year and the technology content doubles every 12 months. Volume products remain in production for only 6–9 months.

Approach

In 1998 Seagate's senior executive team was concerned that business performance was not on par with expectations and capabilities. The quality group was charged with recommending a new model or system with which to run the business. The Six Sigma methodology was selected and launched in 1998 to bring common tools, processes, language and statistical methodologies to Seagate as a means to design and develop robust products and processes. Six Sigma helps Seagate make data-based decisions that maximize customer and shareholder value, thus improving quality and customer satisfaction while providing bottom line savings.

Six Sigma was one of the three key activities seen as essential for Seagate's continuing prosperity. The other two were:

- *Supply chain*: How to respond to demand changes in a timely manner, execute to commitments and provide flexibility to customers.
- *Core teams*: How to manage product development from research not sure what you are saying hereto volume manufacture.

Implementation

Seagate Springtown (which is part of Seagate Recording) started a supply chain project to improve materials management and develop a strategic vendor relationship. The fabrication plan at Springtown introduced the Lean Manufacturing philosophy that recognises WASTE as the primary driver of cycle time and product cost. Very soon a change had taken place at Springtown, and Lean Manufacturing was wholly integrated with the supply chain initiative.

The corporate office at Scotts Valley was rolling out a global Six Sigma deployment programme. The Springtown site followed the Six Sigma training programme and implemented a number of tools and techniques including the Process Map, Sampling Plan, Cause and Effect Analysis and Control Plans, which identified a 'hidden factory'. The less visible defects of this 'hidden factory' included:

- Repeated measurements (in and out)
- Repeated chains (post and pre)
- Transits between manufacturing areas
- Process steps conducted in 'non-standard operating conditions'
- High rework on a process.

Results and learning points

The Six Sigma methodology proved a key enabler for supply chain/Lean Manufacturing and the integrated programme achieved improved process capability and quality as shown by:

- Increased throughput by 31%
- Significant impact on capital expenditure due to increased efficiency of existing equipment
- Lower work in progress
- 80% pass rate on qualifications for vacuum tools (previously 40%).

The main learning points from the Six Sigma programme at Seagate Technology include:

1. Companies using Six Sigma need to learn how to use the metrics to manage – to make appropriate decisions on a holistic basis, avoiding sub-optimisation. This task of integration with the whole of the company's business process is the key.
2. Set aggressive goals – don't make them too easy.
3. Develop a system for tracking 'soft savings'.
4. Develop a common language and encourage its use on a widespread basis early in the programme.
5. Embed the business process within the organization by training all functions – use Green Belt, Black Belt and customized programmes as appropriate.

Tools and techniques used

IPO Diagram, CTQ Tree, Value Stream, Flow Process Chart, Cause and Effect Diagram, Brainstorming.

Case Study 6: FIT SIGMA™ at a SME, Sweden

Background

The Solectron factory in Ostersund, Sweden, where AXE switchboards are manufactured, employs approximately 1000 people. The site was formerly part of the Ericsson Network of core products AB. Solectron as an independent company was experiencing tough competition even at the crest of the 'telecom boom'. With the downturn of the market from 2000, the competition became increasingly fierce. The management were toying with the idea of launching a Six Sigma initiative, but their initial enquiry revealed that they would be set back by at least $1 million if they began a formal Black Belt training programme with the Six Sigma Academy at Scottsdale, Arizona.

Approach

Solectron applied the FIT SIGMA™ methodology – 'Fitness for the purpose' – albeit not consciously under the FIT SIGMA™ label (see Basu and Wright, 2003).

Ericsson, the parent company of previous years, had already embarked upon a Six Sigma initiative. The Black Belt training programme was also in full swing. Solectron decided to send a promising manager to a Black Belt training course via the Ericsson deployment plan.

Implementation

The young manager duly returned to Ostersund with great enthusiasm and applied a preliminary 'base line analysis' rooted in a simple checklist. The results were then presented to the management and a customised programme was drafted. The training programme was extensive, but Solectron relied on the Black Belts from Ericsson and also retained the same consultants as and when required.

Ten members of the top management team attended a 1-day course on Six Sigma, 14 people were trained as 'Black Belts' on a 7-month part-time programme and 20 more attended a 2-day course.

Results and learning points

Six Sigma applications at this factory saved US$0.5 million during the first year of the project. This amounted to about $500 per employee, but was actually closer to a huge $36 000 per employee trained in Six Sigma methods. A modest start in terms of savings per Black Belt perhaps, but the investment was also a fraction of a 'pure play' Six Sigma initiative. More significantly, this customised approach enabled Solectron to have a launch pad to gain a much needed competitive advantage.

The key learning point of Solectron was that it was possible to gain significant benefits by adopting Six Sigma principles customised to the size and capability of the organisation.

Tools and techniques used

IPO, SIPOC, CTQ Tree, Project Charter, Process Map, Histogram, Control Chart, Cause and Effect Diagram, Brainstorming, DMAIC.

Case Study 7: Performance monitoring at Dupont Teijin Films, UK

Background

Dupont Teijin Films is a global polyester films business with manufacturing sites in the United States, Europe and Asia. The company was created following the acquisition of Teijin Films of Japan by Dupont. DTF is a market leader but experiencing tough competition from new entrants. As part of the corporate Six Sigma programme, the Wilton Site of DTF in Middlesbrough, UK, started the deployment plan from 1999.

Approach

The main objectives of the programme included:

- Increased capacity
- Improved material efficiencies
- Cost reduction
- Increased revenue by higher sales volume.

The site project team followed a methodology of 'successful implementation' in three key categories – 'Doing the Right Work' (Process); 'Doing the Work Right' (Efficiency) and 'Creating the Right Environment' (Education and Culture).

Implementation

Within the category of 'Doing the Work Right' the team introduced:

- Input Metrics
- Output Metrics
- Tracking Profile.

The Input Metrics included the number of Black Belts trained and people trained. The Output Metrics covered:

- $ saved
- Number of projects per annum
- Quality index
- CTQ flowdown
- COPQ
- Strategic linking.

The project team followed an internal self-assessment process every quarter based on a 'Do Right Work Checklist' comprising 24 questions in Customer Alignment, Business Alignment, Process Baselining and Project Selection.

Results and learning points

The Six Sigma programme has been running at the Wilton Site since 1998. The typical sources and relative magnitude of savings are given in the following table:

Source	Year 1 (%)	Year 5 (%)	Relative value of savings
Low hanging fruits	40	5	$$
COPQ	40	20	$$$$
KPIs	10	10	$$
CTQ flowdowns	10	30	$$$$
Strategic linking	0	35	$$$$$

The key learning point of DTF at Wilton is that training is key deployment opportunity to influence the hearts and minds of your people. The structured continuous training programme has been the key to sustaining the Six Sigma programme for over 5 years.

Tools and techniques used

IPO Diagram, Project Charter, CTQ Tree, Process Capability Measures, Process Map, Control Chart, Balanced Scorecard, DMAIC, DFSS.

Case Study 8: Total productive maintenance at Nippon Lever, Japan

Background

The Utsunomia plant in Japan was commissioned in 1991 on a greenfield site by Nippon Lever to manufacture household detergents products and plastic bottles for liquid detergents. The factory was experiencing 'teething' problems primarily due to the poor reliability and lack of local support of the imported equipment. Many of the employees were new to factory work.

To improve this situation the company got the help of the Japanese Institute of Plant Maintenance (JIPM), an organisation which is working on TPM (Total Productive Maintenance) with over 800 companies in Japan. TPM has been widely used in Japan, having been developed to support Lean/JIT and TQM. It was considered to be appropriate for the Utsunomiya plant and focuses on machine performance and concentrates on operator training and teamwork.

Approach

A TPM programme was launched at the Utsunomiya plant in July 1992 with the objective of zero losses:

- Zero stoppages
- Zero quality defects
- Zero waste in materials and manpower.

Strong organisational support was provided by the Nippon Lever management in terms of:

- A top management steering team to facilitate implementation by removing obstacles
- A manager to work full time supporting the programme
- One shift per week set aside for TPM work
- Training for managers, leaders and operators involving JIPM video training material.

The programme launch was initiated at a 'kick-off' ceremony in presence of the whole Nippon Lever Board and managers from other company and suppliers' sites.

Implementation

The initial thrust of the programme was the implementation of 'Autonomous Maintenance' following the JIPM's seven steps:

1. Initial clean-up
2. Elimination of contamination
3. Standard setting for operators
4. Skill development for inspection
5. Autonomous inspection
6. Orderliness and tidiness
7. All-out autonomous working.

To implement the seven steps, 'model machines' (those giving the biggest problems) were chosen. This approach helps to develop operators' knowledge of a machine and ensures that work on the model can be used as the standard for work on other machines. It also helps motivation. In that if the worst machine moves to the highest efficiency, this sets the tone for the rest of the process.

The improvements to the machines were made using Kaizen methodology (small incremental improvements), and were carried out by groups of operators under their own guidance. Two means of support were given to operators – a Kaizen budget per line so that small repairs and capital expenses could be agreed without delay and the external JIPM facilitator provided encouragement and experience to workgroups.

Results and learning points

By the end of 1993, substantial benefits were achieved within a year at the Utsunomiya plant including:

- £2.8 million reduction in operating costs
- Reduced need for expensive third party bottles
- Production efficiency increased from 54% to 64% for high speed soap lines and from 63% to 80% for liquid filling lines
- A team of trained, motivated and empowered operators capable of carrying out running maintenance.

The success of the programme at the Utsunomiya plant leads to the introduction of TPM to other two factories of Nippon Liver (Shimizu and Sagamihara). Over the next few years the Corporate Groups of Unilever encouraged all sites outside Japan to implement TPM with remarkable successes achieved particularly in factories in Indonesia, Brazil, Chile, United Kingdom and Germany.

Tools and techniques used

Overall Equipment Effectiveness (OEE), Five S, Five Whys, Kaizen, SMED (*Source*: Leading Manufacturing Practices, Unilever Research & Engineering Division, 1994.)

Case Study 9: Sales and operations planning at GSK, Turkey

Background

GlaxoSmithKline Turkey (GSK Turkey, previously known as Glaxo-Wellcome Turkey) was awarded MRPII 'Class A' certification in 1999 by business education consultants Oliver Wight Europe.

GSK Turkey launched a programme (known as EKIP) in January 1998 to improve company-wide communications and sustain a robust business planning process using MRPII 'best practice' principles.

Approach

Since September 1998, the company has improved and sustained a customer service level at 97% and inventory turnover of around 5.0. The sales turnover in 1998 has increased by 20% in real terms in spite of some supply shortfall from the corporate network in the first half and the adverse economic and political conditions of Turkey. GSK Turkey has been recognised as a major business in the pharmaceutical giant GSK Group and the business plan for 1999 was aiming at a turnover of US$110 million.

Implementation

As part of the MRP II Class A programme, GSK Turkey installed an S&OP process, which is underpinned by a set of business planning meetings at various levels. In spite of the Glaxo Wellcome and Smith Kline Beecham merger and the corporate Lean Sigma initiative, the S&OP process has been continued by the company every month.

Results and learning points

The rigour of the S&OP process, championed by the Managing Director, has helped the company to sustain and improve the business benefits and communication culture especially when they were challenged by a number of initiatives in hand, including:

- Transfer of office
- Rationalisation of factory and warehouse
- Corporate Lean Sigma programme
- Merger of GlaxoWellcome and Smith Kline Beecham.

The key learning point of the GSK Turkey experience is that in order to sustain a competitive performance level, S&OP process is extremely useful in reviewing the operational performance as well the strategic objectives of the company in the midst of major changes.

Tools and techniques used

Brainstorming, Five Whys, Balanced Scorecard, S&OP.

Case Study 10: Six Sigma training at Noranda, Canada

Noranda Inc. is a leading international mining and metals company for copper, zinc, magnesium, aluminium and the recycling of metal. With its headquarters in Toronto, the company employs 17 000 people around the world and its annual turnover in 2000 was $6.5 billion.

Approach

In August 1999, the Board of Noranda decided to embark upon a global Six Sigma project with an initial savings target of $100 million in 2000. There are some specific challenges to overcome. The company business is in a traditional industry with long serving industry. Furthermore, Noranda is a 'de-centralised' company with multiple cultures and languages. Senior executives studied the experiences of other companies (GE, Allied Signal, Dupont, Bombardier and Alcoa) and invited the Six Sigma Academy from Arizona to launch the training deployment programme.

Implementation

The Six Sigma structure at Noranda focused on the training of the following levels:

- Deployment and Project Champions
- Master Black Belts
- Black Belts
- Business Analysts ad Validates
- Process Owners
- Green Belts.

The Six Sigma Academy was intensely involved for the first 3 months of the programme and then Noranda started its own education and training.

Results and learning points

The training accomplishments in 2000 were impressive:

- All 84 top executives followed a 2-day workshop
- 90 Black Belts were certified
- 31 champions were trained
- There were 17 days of Master Black Belt training
- 3000 days of Green Belt training
- More than 3500 days of training in 2000 – and this has continued.

The key learning points from the experience of Noranda include:

- The rigorous deployment of structured learning schedules is key to the success of a company Six Sigma programme.

- After the initial learning by external consultant it is more cost effective and useful in the longer term to build up your own in-house training teams.

Tools and techniques used

IPO Diagram, Project Charter, Process Map, Control Chart, Pareto Diagram, DMAIC, DOE, FMEA, QFD.

Case Study 11: Self-assessment at Jansen-Cilag, UK

Background

Janssen-Cilag is the pharmaceutical arm of the Johnson and Johnson Group with their European Head Office based in High Wycombe, Buckinghamshire. The origins of the company lie as far back as the 1940s, with three companies initially in existence: Ortho Pharmaceutical in the United Kingdom, Cilag Chemie in Switzerland and Janssen Pharmaceutical in Belgium. The merger was completed in 1995 and Janssen-Cilag is now among the top 10 pharmaceutical companies in the world. The company markets prescription medicines for a range of therapeutic areas of gastroenterology, fungal infections, women's health, mental health and neurology.

Approach

The commitment of the company to the values and standards laid out in 'Our Credo' drives management to strive continually for excellence in a number of overlapping areas. Based upon the principles of the Baldridge Award, the Quality Management team of Janssen-Cilag developed a self-assessment process known as 'Signature of Quality (SoQ)'. The process is supported by a checklist on a carefully constructed questionnaire in five interdependent areas:

- Customer Focus
- Innovation
- Personnel and Organisational Leadership
- Exploitation of Enabling Technology
- Environment and Safety.

Implementation

SoQ is managed as a global process from the US office and each site is encouraged to prepare and submit a comprehensive quality report meeting the requirements. The assessment is carried out by specially trained

Quality Auditors and a site may receive a SoQ Award based upon the results of the assessment.

Results and learning points

SoQ has been reported to be successful in Janssen-Cilag as a tool for performing a regular 'health check' and as a foundation for improvement from internal benchmarking.

The key learning from the experience of Jansen-Cilag is that even without a declared goal of achieving external quality awards, the self-assessment process customised from a proven format can be very effective in driving a quality programme.

Tools and techniques used

Brainstorming, Balanced Scorecard, Process Map, MBNQA/EFQM.

Case Study 12: Total quality at Chesebrough-Pond's, Jefferson City

Background

The Jefferson City plant Chesebrough-Pond's was built in 1966 and was managed through the traditional functional structure. Each packing department had its own manager and supervisors responsible for each shift resulting in five reporting layers. Maintenance and industrial engineering were separate departments and production planning was located with accounts in the office area. Raw materials and packaging suppliers were managed by the purchasing department in the head office in Connecticut, 1000 miles away.

With the concentration of the purchasing power of the retail trade in the United States (e.g. Wal-Mart, K-Mart, etc.) the plant had come under considerable pressure to deliver personal products more often, at shorter lead times and at a lower cost. Therefore the plant decided it had to increase flexibility while at the same time reducing operating costs.

Approach

The management of the Jefferson City plant recognised that an organisational change was required to meet the demands of the business and launched its Total Quality programme in early 1990s.

The process started with the writing and adoption of statements on Plant Vision, Plant Mission and Plant Values. These statements were communicated throughout the plant. The other key activities in establishing Total Quality at the plant were:

- Understanding the supply chain
- Process management
- Problem solving skills
- Setting goals and measuring progress
- Training.

Implementation

The Jefferson City plant displayed Plant Vision, Mission and Values by posting them in the permanent (yearly) framed charts in primary production buildings.

The plant was divided into three focused manufacturing units replacing the traditional functional management structure. All employees were referred to as 'Associates' and participated in self-directed work groups. The first line supervisors were redeployed as 'Coaches' of workgroups formed by Associates. Each workgroup owned activities such as supplier contract, production scheduling and small projects.

All Associates in the plant received at least 6 days training per year in Total Quality tools and principles. Topics included Quality characteristics, SPC, GMP, Safety, Scheduling and Waste Analysis. Even supplier employees visited the plant for 4 days training in TQ.

Results and learning points

Within 3 years the plant achieved the satisfying of customer expectations through improved quality and matching production on the basis of demand and lower costs. For example,

First pass quality improved from 93% to 99%
Work-in-progress inventory reduced by 86% and total inventory by 65%.

As a result the plant was recognised as one of its 10 best plants in the United States by *Industry Week* magazine in 1992.
(*Source*: 'Our Search for the Best', *Industry Week*, USA, October 1992.)

Case Study 13: Sustaining Six Sigma by Leadership Forums in Pliva Croatia

Background

Pliva is a generic pharmaceutical company with manufacturing sites in Croatia, Poland and Czech Republic with a broad range of product portfolio including *Sumamed.* During the summer of 2004, Pliva embarked upon a Six Sigma programme called PEP (Pliva Excellence Process) and retained an American consulting company to train Black Belts and initiate projects. Although a good foundation of training for 16 Black Belts underpinned by DMAIC based projects was accomplished, a change of mindset and culture was lacking. In 2005 Performance Excellence Limited was engaged to re-launch the programme. The primary focus of the second wave included the Certification of Black Belts with completed projects, training of additional Black Belts and Green Belts and the management buy-in by Leadership Forums. Here is an account and outcome of one such Forum.

Approach

PEP or Pliva Excellence Process came of age. Since the launch of this Six Sigma initiative in Global Product Supply (GPS) in the summer of 2004, PEP demonstrated tangible results as reflected by the projected savings of US$13.4 million of which $4.6 million has already been realised. Perhaps more significantly PEP opened the door towards a self-supporting sustainable Operational Excellence competence and holistic culture in most business units of GPS, supported by trained 28 Back Belts and 108 Green Belts. Time was right to leverage this success and good practice across and beyond GPS. A most critical success factor for a major change programme such as PEP was the better understanding of the process and commitment of managers. With this backdrop in mind the first PEP Leadership Forum was conducted by Zeljko Brebric, PEP Director, supported by Ron Basu, the retained Consultant and Mentor, and the PEP team.

Implementation

The event took place in Zagreb and was attended by 24 managers from Pharma Chemicals, Local Product Supply Zagreb, Finance, Regulatory Affairs, Business Development, Medical Marketing Services and Human Resources. The primary objectives of the forum were to engage front line managers in understanding Six Sigma/FIT SIGMA™ and PEP and identifying data driven projects.

Zeljko Brebric presented an update of PEP and emphasised that PEP would be a continuous journey of achieving and sustaining operational excellence. The present strength of Black Belts and Green Belts would be strengthened by additional waves of training and the next wave of in-house Black Belt education would soon commence. Ron Basu then explained the quality movement leading to Six Sigma and FIT SIGMA™ and what they meant to Pliva. This was followed by presentations by Ron Basu and interactive discussions on DMAIC methodology, project selection, Six Sigma tools and how to make it happen. It was evident that PEP projects could span all business units of Pliva and not just confined to GPS manufacturing operations.

Managers from Pharma Chemicals and Local Product Supply who had involvement in PEP emphasised the advantage of data driven approach leading to sustainable results. Participants from Regulatory Affairs, Business Development, Medical Marketing Services, Finance and HR shared the potential benefits of extending PEP in their business units but also expressed concerns of possible extra workloads on their resources.

Results and learning points

A Product Manager asked, 'In a simple sentence how would I say to Pliva senior management that PEP is not another fad?' Ron Basu replied, 'PEP has tangible results to show and further savings and improvements will be done by Pliva teams in a Pliva way'. Zeljko Brebric added, 'We also aim to build a team of self-sufficient competent managers based upon BB and GB education'.

(*Source*: *PEP News* No. 5, January 2006, published by Pliva Group, Zagreb.)

Case Study 14: Lean Sigma changes a pharma culture

Background

This case study is an outcome of a MBA dissertation at Henley Management College.

The United Kingdom is a division of multinational pharmaceutical company, which manufactures and distributes medical devices, in vitro diagnostic devices and accessories. This division was formed in 2004 after a merger to become the third largest provider of blood glucose monitoring in the world. The holding company in the United States has operated a succession of policies to improve its financial position. This has culminated in a Lean Sigma implementation in its UK division from 2006.

Implementation

A master (Six Sigma) Black Belt was employed to lead the Lean Sigma initiative called Business Excellence Programme. This initiative was run in parallel with the closure of the original site and consolidation to a purpose-built new facility at Oxfordshire. All of the qualified Black Belts within the Business Excellence team run several projects, approximately four, at any given time, Green Belt training and mentoring. There have been no new Black Belts trained since the initial training. Plans are in place to start a Green Belt upgrading programme during 2008.

A single site case study approach was utilised to determine the success of this implementation and to determine whether the change process lead to significant permanent cultural changes. The effect of governmental regulations on the implementation was also considered. Most previous Lean Sigma initiatives have occurred in unregulated industries.

Data was collected by focused interviews with relevant local management and Lean Sigma trained employees. This data was combined with previously unanalysed data and an internal white paper, which correlated quality and regulatory issues with project initiatives at the UK division.

Results

Analysis demonstrated that the implementation of Lean Sigma into the UK division has not progressed at the expected rate. These results overall indicate that there are issues in several areas:

- Regulatory requirements
- Current IT strategy and knowledge management
- Lean methodology use but less Six Sigma implementation
- Cultural change
- Parallel initiatives.

Governmental regulatory issues ensure that this kind of implementation is slower and more complex than the textbook examples for a standard manufacturing company. Projects have been shown to take approximately twice as long as in a non-regulated industry. The perceived slow implementation of Lean Sigma is creating some dissatisfaction.

It is recognised that issues in the regulatory environment and the many information streams are significant to the pharmaceutical world. Knowledge management strategy is crucial to a successful implementation. Fragmented IT systems are a hindrance explaining why Lean tools are used in preference to Six Sigma, as they rely less heavily on analysis of quantitative data.

Cultural change has been slow but appears to be lasting; progress has been made with other initiatives including Kaizen across the site providing added impetus. The Lean Sigma cultural change programme relies on the site management team to drive a true lasting cultural change, owned by all the employees. One significant aspect of cultural change is the inclusion of suppliers in training and delivering projects. At the time of writing this report there are 65 Green Belt projects of which 22 projects are led by suppliers.

Summary of Part 4

Chapter 11 provides practical guidelines for selecting appropriate tools and techniques and making it all happen in a quality programme like FIT SIGMA™. Many a Six Sigma exercise started with high expectations and looked good on paper. Many an organisation has been impressed by success stories of Six Sigma, but unsure how to start.

The implementation plan shown here will enable any organisation at any stage of a Six Sigma initiative to follow a proven path to success and to sustain benefits. Our implementation plan has nine steps beginning with Management Awareness to the ongoing step of Improve and Sustain. There is no end!

In the spirit of FIT SIGMA™, fit for purpose, this framework can be adjusted and customised to the specific needs of any organisation.

At all stages of the programme it is essential that not only is the Executive Steering Committee and the Torch Bearer kept informed (the Torch Bearer will keep the board informed), but that there is open communication with all members of the organisation, so that everyone is aware of the aims, activities and successes of the programme.

In Chapter 11 case studies are included to offer a practical insight of their applications. We have chosen examples of both company-wide programmes and solutions of problems so that both the application of tools and techniques and the holistic approach of implementation can be illustrated.

Part 4: Questions and exercises

1. In the context of your organisation and the status of quality improvement initiatives, examine the steps of implementation described in Chapter 11 and develop an implementation strategy to move towards Total Quality Management or Six Sigma or FIT SIGMA™ as appropriate.
2. A large multinational FMCG manufacturing organisation has a turnover of £12 billion per annum and operates in 32 countries with approximately 60 000 employees. Both the growth and profitability of the company is declining over the last 5 years. Some of the subsidiary companies have carried out isolated cost effectiveness programmes under different banners, e.g. TQM, TPM, BPR and JIT.

   You have been asked as a Management Consultant, with particular expertise in Six Sigma, to develop a corporate strategy for rolling out operational excellence programmes to all subsidiary companies. Outline a plan to include:
   - Objectives and scope
   - Corporate imperatives
   - Methodology, tools and techniques
   - Estimated costs and benefits
   - Implementation steps
   - Roll-out time plan.
3. Your organisation has been successful in the implementation of a quality programme based on Six Sigma principles. Write a case study of the programme by taking into account:
   - Organisation overview (whether service or manufacturing)
   - Drivers for change
   - How DMAIC/DFSS methodology has been used
   - What kind of tools and techniques have been used and why
   - Costs and benefits
   - Learning points.
4. Risk AB in Sweden is an insurance broker and risk consultancy firm and a wholly owned subsidiary of a stock-tested US group. The company has a turnover of SEK 150 million and employs 100 people. Having read the success stories of GE Capital, the CEO of the company is keen to consider an appropriate service excellence programme based on Six Sigma principles. As an internal consultant, how would you develop an action plan?
5. Medicare plc is a world-class pharmaceutical company of 18 000 employees and £15 billion turnover. The company has successfully implemented a Lean Sigma programme. Develop a strategy for the company to continuously improve and sustain the high performance levels.
6. Infotec is a world leader in computer hardware. Over the last decade the software service side of the business experienced the most significant growth now accounting for 40% of the total revenue. The company experimented with a TQM programme in the early 1990s but it was abandoned

after 2 years. The newly appointed CEO wishes to initiate an EFQM driven quality management programme in the EMEA region of the company. Develop an implementation strategy for the CEO.

7. As a Management Consultant with expertise in Operational Excellence recommend appropriate tools and techniques for the following programmes:
    a. FIT SIGMA™ in a chain of hotels
    b. TQM in a mail order firm
    c. Lean Enterprise in the order processing department of a large telecommunication company
    d. Cost improvement in a road construction company
    e. New process development in a research organisation
    f. Improving packaging effectiveness of a FMCG manufacturing company.

Appendix 1
Management models

If a man does not know to what part he is steering
No wind is favourable to him

—Seneca

Introduction

Fortunately in the real world of business management it is still the ideas and solutions. Not models, that matter. However, it occurred to us that a book dealing with the practical ideas of implementing tools and techniques toward achieving improved quality and operational excellence would be incomplete without the most popular management models. The models do not offer solutions to quality problems but they help to assess a 'big picture' and often reduce the complexities involved.

As a conscious effort of differentiating the models from the problem-solving tools and techniques we have included them in the appendix. The abundance of models on offer is source of bewilderment for many managers and consultants alike. We have drawn up a shortlist of 14 of the most frequently cited management models. The description of each model has the following structure:

- Description
- Application
- Final thoughts

A1: Ansoff's product/market matrix

Description

The product/market matrix of Ansoff (1987) is a sound framework for identifying market growth opportunities. As shown in Figure A1.1, the x-axis shows the dimensions of the product, and the y-axis represents the current and the new market.

There are four generic growth strategies arising from Ansoff's grid. These are:

1. *Current product/Current market*: The strategy for this combination is 'Market Penetration'. Growth will take place through the increase of market share for the current product/market mix.

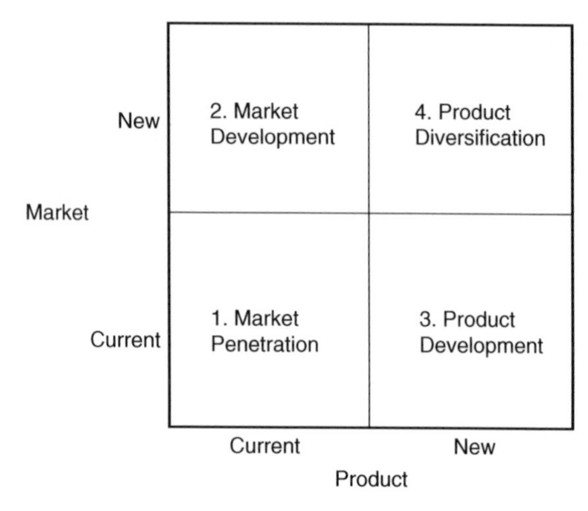

Figure A1.1 Ansoff's matrix (© Ron Basu).

2. *Current product/New market*: In this situation the strategy for growth is 'Market Development'. The pursuit will be for exploring new markets for current products.
3. *New product/Current market*: The strategy of 'Product Development' is followed to replace or to complement the existing products.
4. *New product/New market*: The strategy of 'Product Diversification' is pursued when both the product and market are new in the business.

Ansoff has also identified a number of specific strategies for the diversification quadrant depending on the different market and product combinations:

- *Vertical integration*: When the organisation decides to move into the suppliers' business.
- *Horizontal diversification*: When entirely new products are introduced in the existing market.

Application

Ansoff's model is traditionally in the domain of Marketing Managers for projecting the direction of business growth. In addition to establishing the scope of the product and market mix, the model has been applied in other aspects of shaping the corporate strategy including business growth, competitive advantage, defensive/aggressive technology, synergy and make or buy decisions.

When used in conjunction with the corporate objectives of an organisation, the above five aspects of the model can be applied in the development of the business strategy.

Final thoughts

The model on its own is ineffective to evaluate the best strategy, but it provides an excellent framework for exploring strategic discussions on products and markets. This accounts for the fact that it is still popular with marketing strategists and business school students.

A2: Basu's outsourcing matrix

Description

The matrix of outsourcing strategy by Basu and Wright (1997) is a useful framework for deciding whether to 'make or buy' and selecting the supply partnership. As shown in Figure A1.2, the model uses the core strength (e.g. technology or patent life) as the *x*-axis and the volume of the product as the *y*-axis.

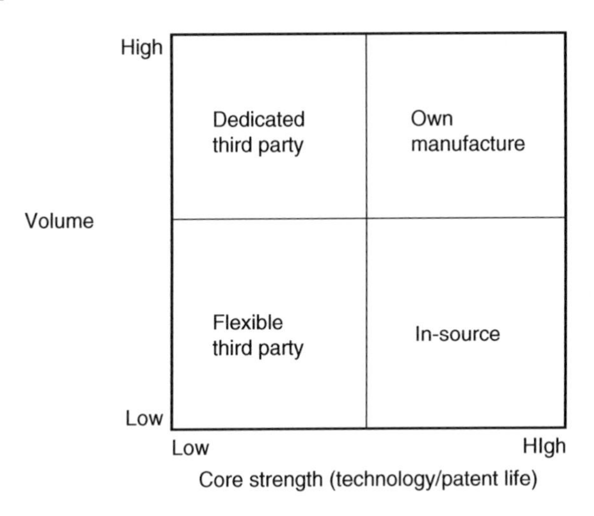

Figure A1.2 Basu's outsourcing matrix (© Ron Basu).

The sourcing strategy of products is considered according to where they appear on the grid as follows:

1. *High technology/High volume*: The products in this quadrant are suitable for 'in-house' manufacture. It will be appropriate to invest to retain the core strength over a longer period.
2. *High technology/Low volume*: When the volume is low the strategy should be to 'in-source'. This means that either the global manufacture of the product is centralised in a single site or the capacity of high technology resources are utilised by bringing in demand from other companies.
3. *Low technology/High volume*: After a period, the technological advantage of a product reduces and it becomes a mere commodity. If the volume is high then a supply partnership is agreed with a 'dedicated third party'.

4. *Low technology/Low volume*: If the volume of the commodity products is relatively low then the product is manufactured by more than one-third party supplier and it does not require a longer-term partnership agreement.

Application

The reduction of tariffs, free market agreements, improved logistics and e-commerce as well as cost effective manufacturing capabilities in the so-called 'third world' countries have created an explosion of outsourcing operations. Few corporations are carrying out all manufacturing operations in-house. The trend is also similar in the service industry and public sectors.

This model has extensive application, especially in larger organisations, by varying the components of the core strength. We have shown only two components, viz. technology and patent life. The matrix can only provide a strategic pointer which should be further evaluated by external factors (e.g. PESTLE analysis) and internal factors (e.g. Financial Appraisal and Risk Analysis). Other dimensions such as quality, responsiveness, dependability, flexibility and innovation are also analysed to establish the outsourcing strategy.

Final thoughts

Basu's outsourcing matrix is a sound framework for identifying the options of outsourcing and categorising them for further analysis. Like other analytical grids it also suffers from the limitation of oversimplification.

A3: The BCG matrix

The Boston Consulting Group (BCG) matrix is one of the early two-dimensional models for analysing and prioritising a product portfolio. A key assumption made by the BCG is that a company should have a portfolio of products that generates both high growth and low growth in a market place.

As shown in Figure A1.3, the matrix contains two dimensions, market share and market growth and creates four categories of product in the portfolio, 'Stars', 'Cash Cows', 'Wildcats' (Question Marks) and 'Dogs'.

1. *High growth/High share*: The products enjoy a relatively high market share in strongly growing markets. The 'Star' products are profitable and good candidates for further investment.
2. *Low growth/High share*: Products in this category are 'Cash Cows' where the market is no longer growing but the market share is still high. The products are profitable but no investment is necessary to sustain the market share.
3. *High growth/Low share*: For 'Wildcats' (or Question Marks) the market share is low but the demand in the market is growing rapidly. The products are in this category and investment in the future may create big profits though this is not guaranteed.

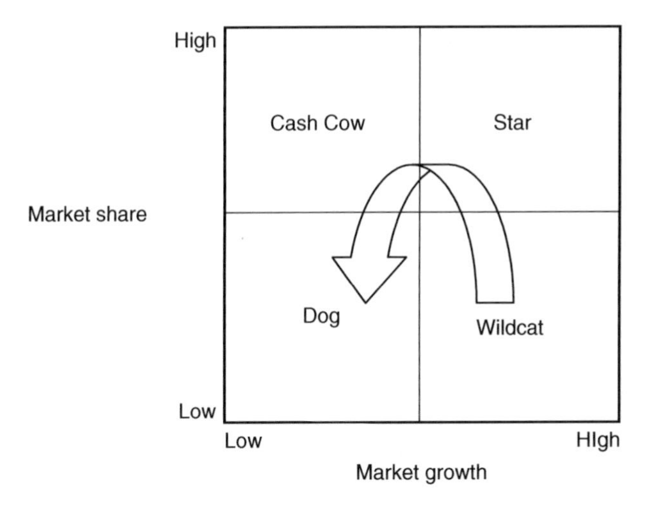

Figure A1.3 The BCG matrix (© Ron Basu).

4. *Low growth/Low share*: When both the market share and its growth are low the product becomes a 'Dog'. The products become unprofitable and are usually divested or discontinued.

Application

Since its inception in the early 1970s, larger companies with a range of products have applied the BCG matrix as a broad framework for allocating resources among different business units. The matrix draws attention to the cashflow, investment characteristics and needs of an organisation's various divisions. The divisions or products groups evolve over time. Wildcats become Stars, Stars become Cash Cows and Cash Cows become Dogs in an on-going counter-clockwise motion.

In spite of its popularity the model has its own limitations, e.g.:

- Analysing every business as either Stars, Cash Cow, Wildcat or Dog is an oversimplification.
- Market growth rate and relative market share are only two factors of competitive advantage. The link between market share and profitability is also questionable.
- The framework assumes that each business division is independent of the others. In some businesses, a 'Dog' may be helping other units as a corporate strategy.

Final thoughts

The BCG matrix has been overused and is not free from flaws. However, it is a useful tool in shaping the direction of the product portfolio, but it should not be employed as the sole means of determining investment strategies for products.

A4: Belbin's team roles

Belbin (1985) defined a team role as 'a tendency to behave, contribute and interrelate with others in a particular way'. Based on his research of over 9 years on the behaviours of managers all over the world, Belbin defined nine team roles divided into three groups (see Figure A1.4). These are:

a. *Action-oriented roles*: Shaper, Implementer and Finisher
b. *People-oriented roles*: Co-ordinator, Teamworker and Resource Investigator
c. *Cerebral roles*: Plant, Monitor and Specialist

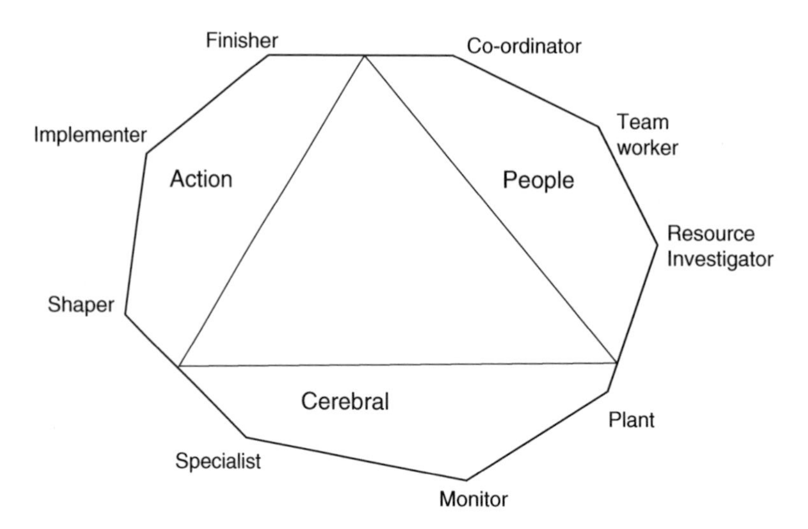

Figure A1.4 Belbin's team roles (© Ron Basu).

The characteristics of each role have been defined as follows:

The Shaper is challenging, dynamic and thrives on pressure. He or she has the drive and courage to overcome obstacles. The Shaper is prone to provocation and is likely to offend people's feelings.

The Implementer is disciplined, reliable, conservative and efficient. He or she turns ideas into practical actions. The Implementer is somewhat inflexible and slow to respond to change.

The Finisher or Completer is painstaking, conscientious and anxious. The Finisher delivers on time. He or she is inclined to worry unduly and is reluctant to delegate.

The next category of worker is the Co-ordinator who is mature, confident and a good chairperson. This individual clarifies goals, promotes decision-making and delegates well. However, the Co-ordinator can often be seen as manipulative.

By contrast, the Teamworker is co-operative, perceptive and diplomatic. He or she listens, balances and averts friction. However, this type of person is indecisive in crunch situations.

An animated, passionate 'people person', the Resource Investigator is extrovert, enthusiastic and communicative. He or she explores opportunities and develops helpful contacts. Although over-optimistic, those belonging to this section tend to lose interest once initial enthusiasm has waned.

The Plant is also known as the Creator or the Inventor. This person is creative and imaginative and brilliant at time. The Plant ignores anything incidental and is too preoccupied to communicate effectively.

Sober, discerning and strategic, the Monitor evaluates options ad judges accurately, but lacks drive and the ability to inspire others.

Finally, the Specialist is a single-minded, dedicated self-starter. He or she provides rare knowledge, but contributes on a narrow front and dwells on technicalities.

Application

With the support of appropriate assessment, Belbin's team roles have been applied in leadership, people development and project team building activities.

There are different forms of assessment:

- Self-assessment or psychometric profiling by answering multiple choice questions.
- Assessment by mentor or line manager during performance appraisal.
- Team assessment or 360° appraisal when members grade each other.

The analysis of Belbin's team roles is particularly useful when an assignment requires the composition of a team or procurement of members of specific skills and combinations of roles. Belbin suggests that team members with complementary roles are more successful.

Final thoughts

The model is conceptually rational but in practice often encourages debates. The objective basis of assessment is questionable. A team, based on this model, may look good on paper but fail to function properly in practice. It also contains too many (e.g. nine) categories.

A5: Economic value added

Description

Economic value added (EVA) calculates the estimate of the intrinsic worth of a company by looking at the present value of all future expected free cash flows in the organisation. This is simpler than it sounds. For example, if a company's capital is $100 million, including debt, shareholders' equity and the cost of using that capital (i.e. interest on debt) is $10 million a year, the company will add economic value for its shareholders only when its projects are more than $10 million a year.

EVA was developed by Stern Stewart & Company who compiles annual performance rankings of large publicly owned companies. EVA is also a measure of the surplus value created in an investment Stern Stewart & Co (1997).

EVA accounts for the cost of doing business by deriving a capital charge. The calculation for EVA is simply

$$\text{EVA} = \text{OPBT} - \text{T} - (\text{TCE} \times \text{WACC})$$

where

OPBT = Operation profit before tax

T = Tax

TCE = Total capital employed

WACC = Weighted average cost of capital

A positive EVA indicate that the company has created value. Often companies become so focused on earnings that they lose sight of the cost of generating those earnings. EVA has become a popular tool for being linked with the executive bonus. It is useful to note that EVA and ROI are closely related as:

$$\text{EVA} = (\text{ROI} - \text{WACC}) \times \text{TCE}$$

where ROI = Return on Investment

Thus as long as ROI is more than WACC the company will add value.

Application

EVA is a popular financial performance measure in the current business environment. It has been applied in many areas of business purposes including:

- Setting targets
- Determining bonuses for managers
- Communicating with investors
- Valuation during acquisition or merger

EVA takes into account the opportunity cost of working capital and relates it to the ROI. Thus it is more meaningful than the more traditional measures such as Earning per Share.

As shown in Figure A1.5, Stern suggests four stages for successful EVA application.

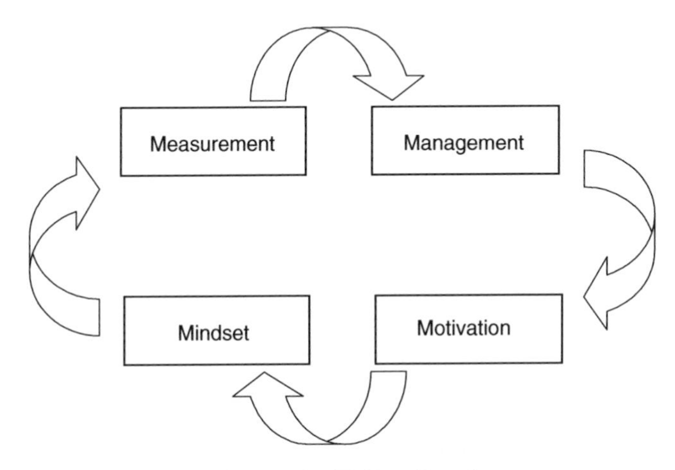

Figure A1.5 EVA application cycle (© Ron Basu).

First establish the rules to convert accounting data to form an EVA calculation. Then track EVA results unit by unit on a monthly basis to complete the requirements of measurement.

With regard to Management, budgeting and planning techniques are adjusted to reconcile with EVA measures. By linking salary incentives with EVA, managers are motivated to increase the shareholder value. Thus individuals are encouraged to think and work in the interest of the shareholders. It has the potential to change the mindset and thus transform the company culture.

Market Value Added (MVA) is the complementary measure to EVA devised by Stern Stewart & Company. MVA is the difference between a company's fair market value (in the stock price) and the economic book value of capital employed.

Final thoughts

EVA is a powerful measure with potentials to motivate managers to act and think like they own or are directors of the business. However, its merits should be weighed against two important drawbacks. First there are several anomalies in trying to calculate the true cost of working capital. Secondly, managers are often discouraged when the share price of the company falls even after showing a theoretical growth in EVA.

A6: The Fifth Discipline

Description

Peter Senge (1994) described five key disciplines that constitute the main elements of a learning organisation. These are:

1. Personal Mastery
2. Mental Models

3. Shared Vision
4. Team Learning
5. Systems Thinking

Senge calls Systems Thinking as the 'Fifth Discipline' because it makes other disciplines work together.

Personal Mastery relates to the ability to focus energy, personal vision and to see reality clearly and objectively.

Mental Models determine how we can see and interact with the world. This discipline helps to open up alternative perspectives and insight and balances enquiry and advocacy.

Shared Vision emerges from personal visions fostering the commitment of the group. The more people share the vision, the more likely it is that the goal becomes achievable.

Team Learning relates to the discipline of learning together. It distinguishes between dialogue (i.e. free exploration of complex issues) and discussion (i.e. the presentation of different opinions in search of the best solution).

The Fifth Discipline, Systems Thinking, provides a way of understanding practical issues to see the interrelationship between processes and cause and effect chains. Systems Thinking relies on 'feedback' which refers to how actions can cause or counteract each other. Senge also suggests that we speak of 'archetypes' when actions influence each other in a close loop. System 'archetypes' are basic and understandable cycles that systems run through.

Application

Senge's Fifth Discipline attracted both enthusiasts and critics. Its critics find the principles philosophical and theoretical. However, the enthusiasts are impressed by Senge's definition of a learning organisation underpinned by the five disciplines. It is worth noting that both critics and enthusiasts have requested tips for the daunting task of putting the Fifth Discipline into practice.

As shown in Figure A1.6, each discipline has different aspects of Essence, Principle and Practice.

The Fifth Discipline brings the concept of 'learning organisations' where people continually expand their capacity to recreate the results they truly desire, where collective aspiration is set free and where people are continually finding out how to learn together.

The disciplines, at least in principle, can be applied beyond business organisations. For example, these are also applicable to family, community groups and to all society.

Final thoughts

The Fifth Principle is likely to be perceived as an 'ivory tower' approach to solving practical problems. It is like someone asking how to start a car and receiving an answer in terms of the laws of physics. However, the concept of

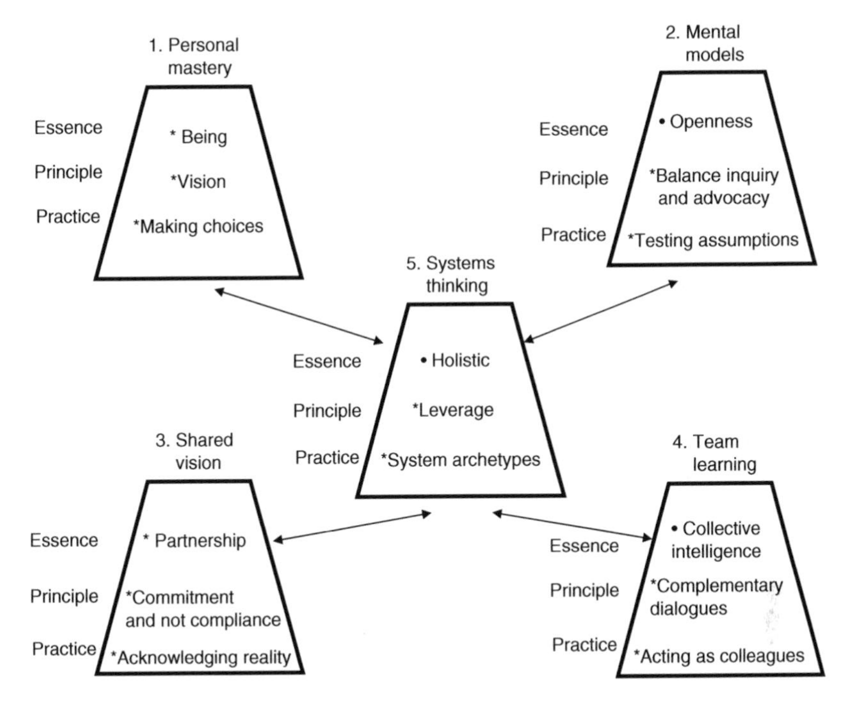

Figure A1.6 The Fifth Discipline (© Ron Basu).

'learning organisations' is invaluable in the current environment of collaborative management and supplier partnership.

A7: The McKinsey 7-S framework

Description

The McKinsey 7-S framework was originally developed by Pascale (1990) as a way of thinking more effectively about organising a company strategy. McKinsey Management Consultants divided the seven organisational elements into what they called 'hard' and 'soft' elements are as follows (see Figure A1.7).

The first three Ss are the 'Hard' or tangible elements, i.e. Strategy, Schedule and Systems.

Strategy relates to the informed choices an organisation makes to achieve its business objectives, such as prioritising the investment projects.

Structure refers to the organisational hierarchy and division of functions and activities.

Systems concerns the primary and secondary processes that the organisation employs to get operations completed, such as manufacturing resource planning, supplier relationship management, etc.

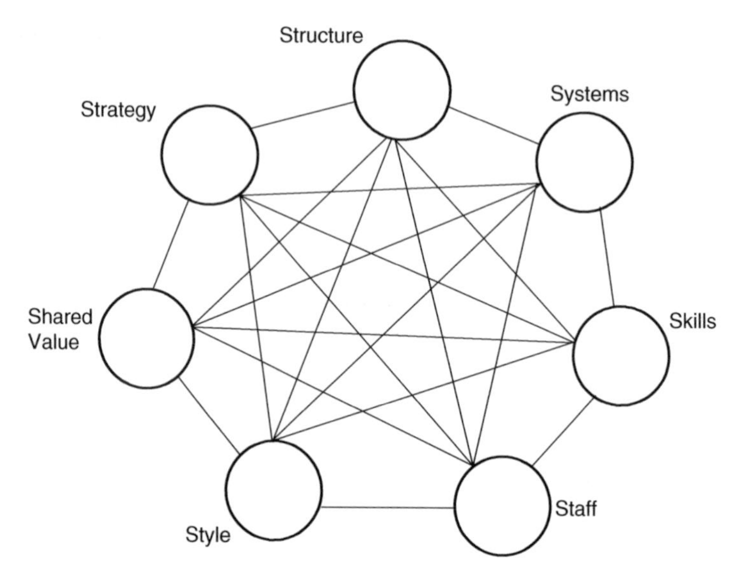

Figure A1.7 The McKinsey 7-S framework (© Ron Basu).

'Soft' elements comprise a further four concepts beginning with 'S': Shared Values, Style, Staff and Skills.

The core beliefs and expectations that employees have about their organisation can be termed as its Shared Values.

Style relates to the unwritten but visible behaviour of the management in conducting business.

Staff is of course the people employed in the organisation.

The collective capability of an organisation is known as its Skills. These are independent of individuals but dependent on the other six Ss.

Application

The 7-S framework has been used by consultants as a good checklist to analyse key element of an organisation. At a more sophisticated level the framework can be used for assessing the viability of a strategic plan. In this case, the 7-Ss are considered as 'compasses' and each element is examined to assess whether or not they are all pointing in the same direction.

By constructing a matrix, the potential conflict between each element can be identified for further analysis. The decisions are made as to how to effect changes in the strategy in order to balance the conflicts. Pascale (1990) argues that smarter companies use conflicts to their advantage. They power the engine of root cause analysis in pursuit of excellence.

Final thoughts

The use of the 7-S Framework as a checklist is useful, but is suffers the risk of being a dressed-down application. The seven elements, especially those in the 'Soft' category, are broad based and lack a detailed a checklist to make them more specific.

A8: Kano's satisfaction model

Description

The Kano model of customer satisfaction demonstrates that blindly fulfilling customer requirements has a risk if the supplier is not aware of different types of customer requirements. The model divides product attributes into three categories:

- Threshold or basic needs
- Performance needs
- Excitement needs

As shown in Figure A1.8, the model provides the above three customer requirements in two dimensions. The first dimension measures the degree to which the customer requirement is fulfilled. The second dimension (y-axis) is the customer's subjective response of satisfaction to the first dimension.

Threshold (or basic) attributes are the expected attributes of a product and do not provide an opportunity for product differentiation. Examples of this category are car brakes or a telephone dial tone.

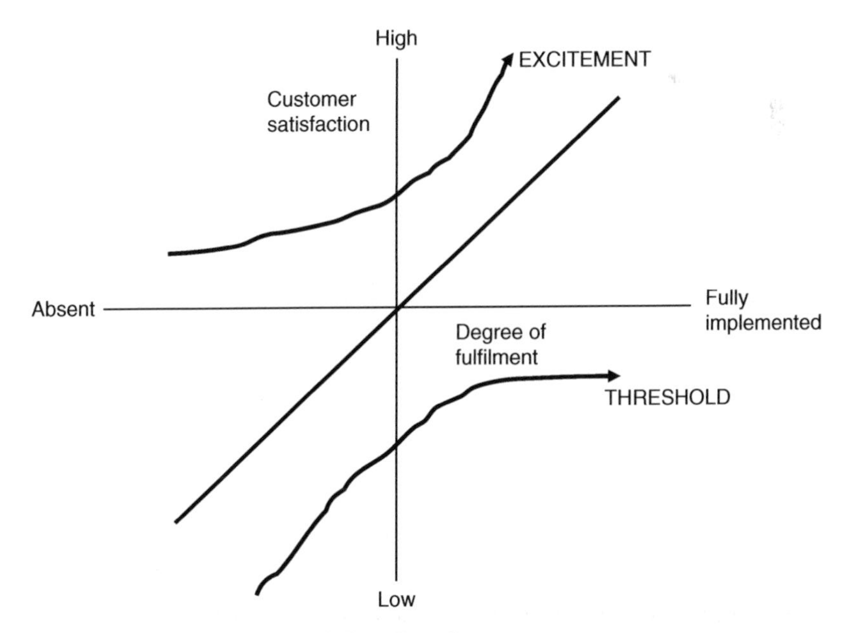

Figure A1.8 Kano model (© Ron Basu).

Performance attributes are those for which more is better and are often referred to as 'fundamental' quality. Customers overtly state these needs. Examples are price, delivery and performance.

Excitement attributes are 'latent' needs but can result in high levels of customer satisfaction. Customers are unaware of these needs until they experience them. Examples are cup holders and Post-it™ notes.

Application

The information obtained from the analysis of the Kano model especially related to performance and excitement attributes, provides valuable input to the Quality Function Deployment (QFD) process.

The basic tool has two sided questions; one is asked in a positive case and the other is asked in a negative scenario, e.g.:

- Rate your satisfaction if the product has the attribute.
- Rate your satisfaction if the product did not have the attribute.

The customer's response should be one of the following:

a. Satisfied
b. Neutral
c. Dissatisfied
d. Don't care

Surveys are then tabulated. Basic attributes generally receive 'Neutral' to Question 1 and the 'Dissatisfied' response to Question 2. Consideration should be given to a 'Don't care' response as they will not increase customer satisfaction. Prioritisation of matrices can be useful in determining which excitement attributes would maximise customer satisfaction.

Final thoughts

The Kano Model is useful in identifying customer needs and analysing competitive products. The model establishes that to be competitive, products and services must execute flawlessly all three attributes of quality – basic, performance and excitement. Exceeding customer's performance expectations creates a competitive advantage.

A9: Mintzberg's organisational configuration

According to Mintzberg (1990) an organisation's structure is determined by the environmental variety which in turn is determined by both environmental complexity and the pace of change. He defines four basic types of organisational form (see Figure A1.9) which are associated with four combinations of complexity and change.

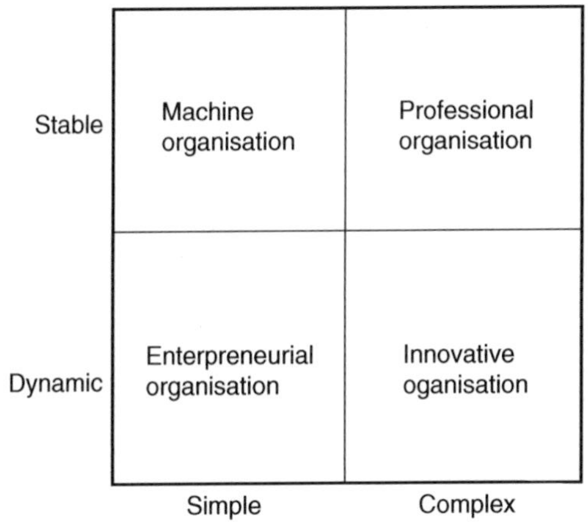

	Simple	Complex
Stable	Machine organisation	Professional organisation
Dynamic	Enterpreneurial organisation	Innovative oganisation

Figure A1.9 Mintzberg's taxonomy (© Ron Basu).

Mintzberg also defines five basic organisational subunits (see Figure A1.10) to help explain each of the four organisational forms.

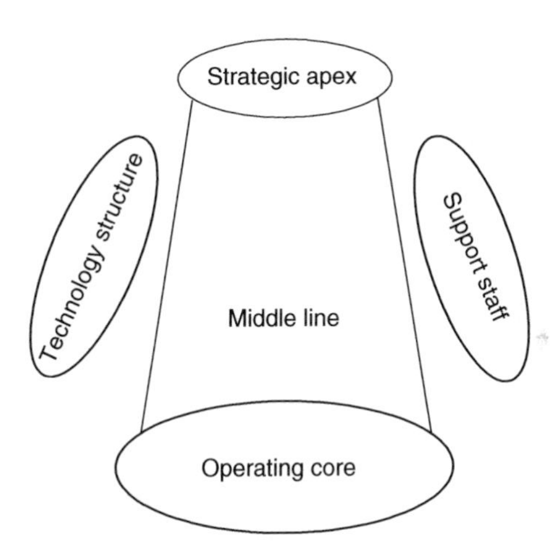

Figure A1.10 Mintzberg's organisational subunits (© Ron Basu).

The typical examples of the functional roles related to each subunit are shown in the context of a manufacturing company (Table A1.1).

The four organisational configurations in Mintzberg's scheme depend on distinctive mechanisms for co-ordination. The power is divided throughout the organisation by different forms of decentralisation according to the configuration. The relevant characteristics of each configuration are summarised in Table A1.2.

In his revised writing, Mintzberg introduced other variants of configuration, such as Diversified Organisation, Missionary Organisation and Political Organisation.

Table A1.1 Functional roles

Subunit	Roles
Strategic Apex	Board of Directors
Technology Structure	Strategic planning, Engineering, Systems analysis
Support Staff	Human resources, Payroll, Maintenance, Catering
Middle Line	VP operations, VP marketing, Sales managers
Operating Core	Machine operators, Assemblers, Warehouse operators, Sales persons

Table A1.2 Configuration characteristics

Configuration	Co-ordinating mechanism	Key subunit	Decentralisation
Entrepreneurial Organisation	Direct supervision	Strategic Apex	Vertical and horizontal centralisation
Machine Organisation	Standardisation of work	Technology Structure	Limited horizontal decentralisation
Professional Organisation	Standardisation of skills	Operating Core	Horizontal decentralisation
Innovative Organisation	Mutual adjustment	Support Staff	Selected decentralisation

Application

Mintzberg's taxonomy offers a number of practical advantages:

- It focuses the way in which power is divided throughout the organisation.
- Having determined the category of an organisation, you can also determine what changes are needed to make it internally consistent.
- It enables the analysis of the co-ordinating mechanism most appropriate for the current configuration and the need for changes.

When an organisation lacks the driving subunit and a prime co-ordinating mechanism, it is more likely to become a political organisation. There is no defined co-ordinating mechanism or a predominant subunit in a political organisation, and thus subunits fight to fill the power vacuum.

In a practical environment, different configurations may form the whole organisation. For example, the University is a Professional Organisation but support units may be composed of other configurations. Support subunits that perform routine functions may have a Machine Organisation, but technocratic subunits may be managed as an innovative Organisation or a Professional Organisation.

Final thoughts

Doubtless, Mintzberg's model helps us to understand the relationship between the nature of an organisation and its co-ordinating mechanism. With four basic criteria, it is difficult to 'fit' an organisation to the typical configuration. Most of the larger organisations are hybrids of multiple configurations.

A10: Porter's competitive advantage

Description

The fundamental basis of above average profitability of a company within its industry determines its competitive advantage. According to Michael Porter (1985), there are three generic strategies for achieving above average performance in industry:

- Cost Leadership
- Differentiation
- Focus

As shown in Figure A1.11, the focus strategy has two variants, cost focus and differentiation focus.

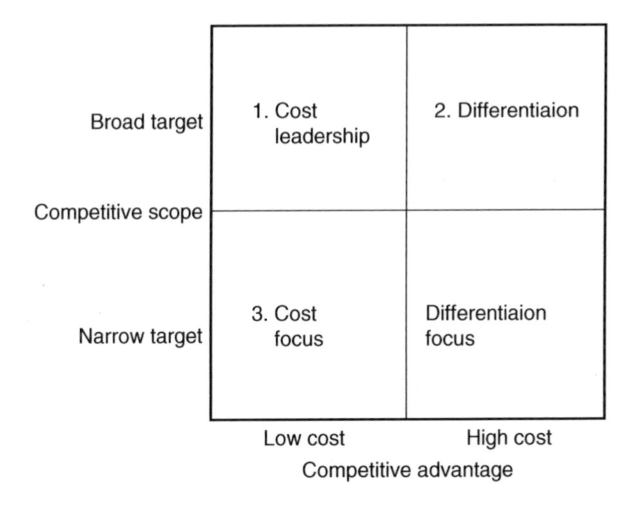

Figure A1.11 Porter's competitive advantage matrix (© Ron Basu).

Cost Leadership: A company in cost leadership sets out to become a low cost producer in its industry. A low cost producer must utilise all sources of cost advantage, such as economies of scale, asset utilisation, preferential procurement of raw materials and other factors to improve efficiency.

Differentiation: A company in a differentiation strategy aims to be unique in its industry among some features that are widely valued by customers regardless of cost. Differentiation requires higher investment in research, design and customer service.

Focus: The generic strategy of focus selects a segment in the industry and customises its strategy to satisfy the target segment. In cost focus a company aims toward a cost advantage in the target segment while in differentiation focus a company seeks differentiation in its target segment.

Application

The model is so broad based that it is logical to assume that a company has no option but to choose one of the three generic strategies. There will always be a competitor in the industry that aims to be either cheaper or better differentiated than your company.

There is likely to be one cost leader in any industry but it is possible to have multiple differentiators. Cost has only one dimension while differentiation can be achieved in several ways.

The initial step of applying this model is to determine the position of a company in the matrix. The next, and probably the most difficult, step is to map the main competitors in the grid. Having established the current relative position of the company appropriate steps can be taken either to maintain or improve it.

Final thoughts

Porter's generic strategy model reduces the guess work in strategic planning but it only provides an initial pointer. Further analysis with the aid of other tools and techniques is required to formulate a practical and appropriate strategy for the company.

A11: Porter's five forces

Description

Michael Porter's (1985) model for competitive analysis relates to five different forces (see Figure A1.12).

1. *New entrants*: Threat of entry from other organisations. Example: Online banks challenging high street banks.
2. *Substitutes*: Availability and competition from substitute products. Example: e-mail as a threat to fax machines.
3. *Buyers*: Bargaining power of buyers. Example: Supermarkets' increasing power related to suppliers.
4. *Suppliers*: Bargaining power of suppliers. Example: Trained specialist required by an industry.
5. *Existing competitors*: Rivalry among existing competitors. Example: Vodafone and T-Mobile jockeying for position in the 3G market.

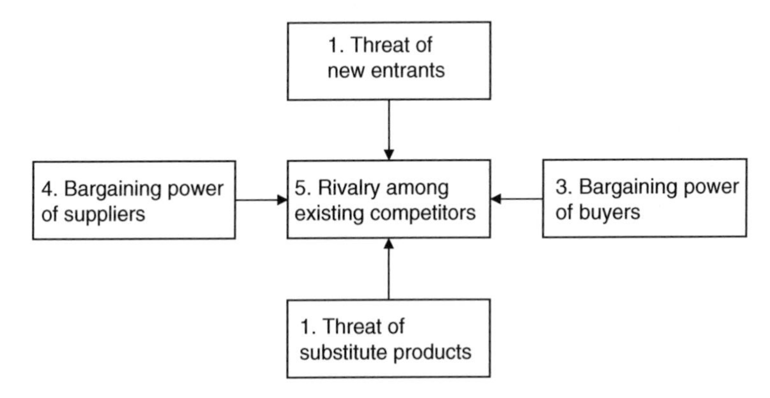

Figure A1.12 Porter's Five Forces (© Ron Basu).

Application

The model emphasises the external forces of competition and how these can be countered by the company. It implies the danger inherent in focusing on your immediate competitors.

The elements involved with each force are shown below to prepare for competition.

New entrants: Examine the entry barriers for new entrants including:

- Economies of scale
- Brand identity
- Capital requirements
- Switching cost
- Access to distribution

Substitute: Analyse the determinants of substitution threats including:

- Relative price performance
- Switching cost
- Buyer's inclination to substitute

Buyer: Examine to what extent buyers can bargain by considering:

- Buyer volume
- Buyer information
- Decision-maker's incentives
- Switching costs
- Differentiated products
- Impact on quality/performance

Supplier: Competitive forces from suppliers mirror those of buyers. Examine the determinants of supplier power including:

- Differentiation of inputs
- Supplier volume

- Substitute inputs
- Switching cost
- Forward integration to your customers

Existing competitors: Analyse the rivalry determinants related to existing competitors. These factors could include:

- Industry growth
- Diversity of competitors
- Fixed cost and asset bases
- Switching cost
- Brand identity
- Exit barriers

Final thoughts

Porter's model of the Five Forces has been most widely used in strategic analysis and business schools. However, the model is more useful for developing a reactive strategy. It is weak in developing a pro-active strategy building upon the core strengths of a company.

A12: Porter's Value Chain

Description

According to Michael Porter (1985), the competitive advantage of a company can be assessed only by seeing the company as a total system. This 'total system' comprises both primary and secondary activities (see Figure A1.13).

The primary activities are made up of inputs, transformation processes and outputs. There are five primary activities, such as:

- Inbound logistics
- Operations

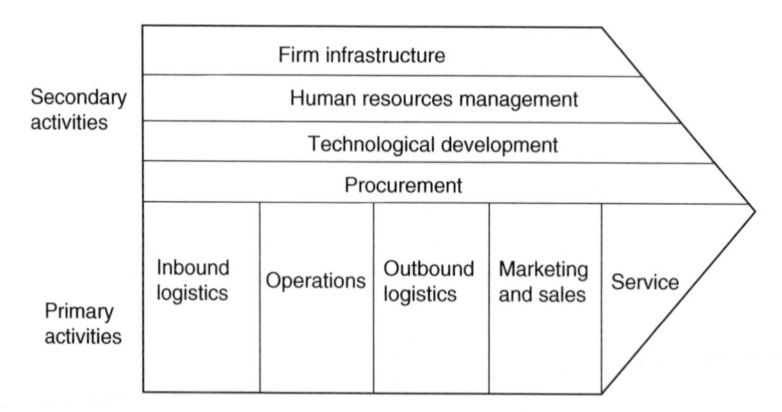

Figure A1.13 Porter's Value Chain (© Ron Basu).

- Outbound logistics
- Marketing and sales
- Services

The secondary activities are the activities that support the primary activities. The four secondary activities are:

- Infrastructure
- Human resources management
- Technological development
- Procurement

Inbound logistics involve relationships with suppliers and include all activities required to receive, store, list and group input materials. Example: Warehousing.

Operations are all activities to transform input into outputs. Example: Assembly.

Outbound logistics include all activities required to collect, store and distribute the output. Example: Distribution management.

Marketing and sales are all activities that convince customers to purchase company products. Example: Advertising.

Service includes all activities required to maintain products after sale. Example: Providing spare parts.

Infrastructure ties companies' various parts together. Example: General management.

Human resources management includes the recruitment, education, compensating and, if necessary, dismissing of employees. Example: Training department.

Procurement is the acquisition of inputs or resources for the company. Example: Purchasing.

Technology development relates to the hardware, software, procedures and technical knowledge brought to transform inputs into outputs. Example: R&D.

Application

Value Chain was originally developed to support general strategic analysis. According to Porter, an organisation may gain a competitive advantage by managing its Value Chain more effectively than its competitors.

As many of the Value Chain activities are independent, the Value Chain model can also be used to represent the linkages between activities. For example, a manufacturing organisation is likely to have important external linkages between its sales and marketing activities and the procurement activity of its customer.

The Value Chain model has been used as a versatile tool for top level analysis, such as a visualisation of a company or a competitor, comparison of competitive strengths and as a potential match for mergers and acquisitions or strategic alliances.

With the emergence of e-business, the application of the Value Chain has been extended to the concept of a 'virtual' Value Chain.

Final thoughts

Like other examples of Porter's models, the Value Chain is an excellent tool for top level concepts. For a detailed analysis leading to a solution, other appropriate tools and techniques are necessary. The Value Chain should not be confused with the detailed features of Value Stream Mapping (see I6 in Chapter 7).

A13: Turner's Project Goals and Methods Matrix

Description

The Goals and Methods Matrix was developed by Turner and Cochrane (1993). It identifies four types of projects according to the clarity of goals and methods. As shown in Figure A1.14, the x-axis shows the Project Goals and the y-axis illustrates the Work Methods. The four types of projects are:

a. *Type 1*: Projects with well defined goals and well defined methods of achieving them are in this category. Example: Engineering projects.
b. *Type 2*: Projects in this category are those with well defined goals but poorly defined work methods. Example: Product Development projects.
c. *Type 3*: Projects with poorly defined goals but well defined methods are in Type 3. Example: Information Systems projects.
d. *Type 4*: This type has both poorly defined goals and work methods. Example: Research projects.

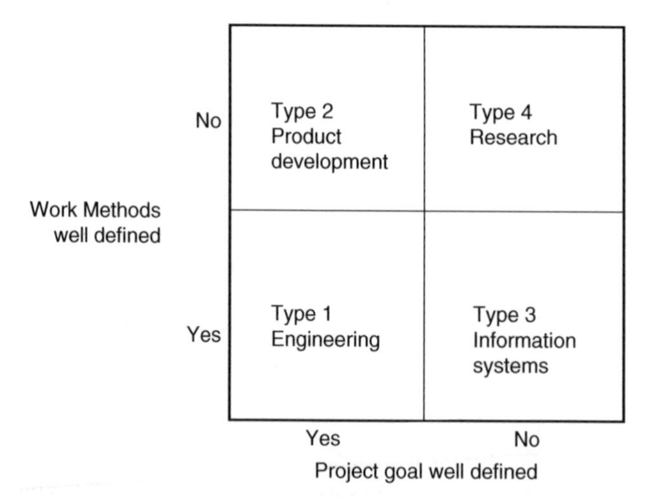

Figure A1.14 Method and Goal Matrix (© Ron Basu).

Application

It is useful to identify the type of project with the aid of this model. Type 1 projects with well defined goals and methods have the best chance of success. On the other hand, Type 4 projects with only ill defined goals and methods have the greatest likelihood of failure. The aim for all projects should be to define the goals first, followed by the work method. An organisational change project may start as a Type 4 project but can move gradually to a Type 1 project for a higher success rate.

Turner suggests that the identification of the project type can help to define the leadership requirement of a project manager. For example, the role of a project manager is as a conductor in Type 1, a coach in Type 2, a master craftsman in Type 3 and an eagle in Type 4.

Final thoughts

The Goals and Methods Matrix is a useful tool to identify the approach to a project start. However, it is important to emphasise that a clear definition of objectives and scope is essential for all types of project.

A14: Wild's taxonomy of systems structures

Description

Ray Wild (2002) introduces the taxonomy of systems structures so that the high level feasibility of an operating system can be examined with regard to its desirable objectives. Wild's taxonomy comprises seven structures (see Figure A1.15), four for manufacturing and supply operations and three for service operations.

Using simple systems terminology as shown in Figure A1.16, all operating systems may be seen to comprise input, process and outputs and either inventory or customer queue (except DOD).

Following Wild's labels of systems structures they are briefly described below:

- *SOS*: Manufacture or supply from stock and the customer is served from an inventory of finished goods. Example: drugs manufacture.
- *DOS*: Manufacture or supply from source and the customer is served from a stock of finished goods. Example: off-shore oil production.
- *SOD*: Manufacture or supply from stock and goods are made to the customer's order. Example: retail shop.
- *DOD*: Manufacture and supply from source direct to customer and no input or output inventories are held. Example: builder.
- *SCO*: Service from stocked input resources but no customer queuing exists. Example: Fire Service.

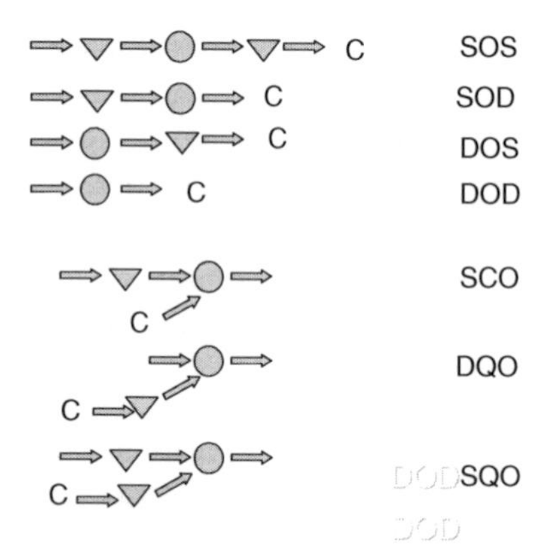

Figure A1.15 Wild's taxonomy of systems structures. There are no 'standard' techniques for mapping operations, but most rely on similar symbols. The technique used by Slack et al. for instance in chapter 5 of their *Operations Management* Book (Financial Times Publishing, 1998) is similar to Wild's. The advantage of Wild's technique is that it can be used at a higher level, to represent capacity buffers rather than just stocks.

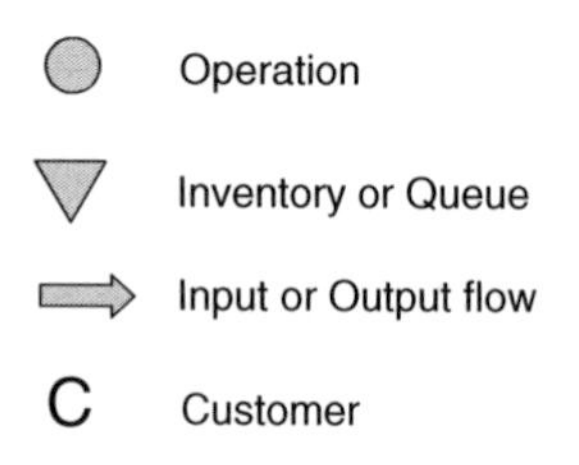

Figure A1.16 Symbols in Wild's taxonomy (© Ron Basu).

- *DQO*: Service from source without dedicated input resources but customer queuing exists. Example: Bus Service.
- *SQO*: Service from stocked input resources and customer queuing exists. Example: checking in for Economy air travellers.

Application

These basic systems structures are useful to describe an operating system for both the total operation and its subsystems. The system structure is then examined for the desired objectives of the operation. For example, if the objective is to deliver a high level of customer service then it is more appropriate to have a structure with buffers offered by stocks (S) or queues (Q).

Customers exert some 'push' in the service models (SCO, DQO and SQO) while in manufacturing and supply models (SOS, DOS, SOD and DOD) customers 'pull' the system.

The choice of a system structure in both manufacturing and service will depend on its appropriateness, feasibility and desirability. If the structure is not desirable then it can be changed with the appropriate change in policy. For example, the original structure SQO for all customers can be changed to SCO for priority customers and SQO or DQO for non-priority customers.

Final thoughts

Wild's taxonomy works well for the overall operation. For a more complex operation it does not pinpoint the ineffectiveness of the system. Other tools, such as the Flow Diagram or Process Mapping, should be more appropriate to describe or analyse a multi-stage complex operation.

Appendix 2
Introduction to basic statistics

Statistics

Statistics is the art and science of using numerical facts and figures. In Wikipedia, statistics is defined as a mathematical science pertaining to the collection, analysis, interpretation or explanation, and presentation of data.

There are three kinds of statistics:

- Descriptive statistics
- Inferential statistics
- Causal modelling

- Descriptive statistics primarily deals with the description and interpretation of data by figures, graphs and charts.
- Inferential statistics is the science of making decisions in face of uncertainty by using techniques such as sampling and probability.
- Causal modelling is a part of inferential statistics. It is aimed at advancing reasonable hypotheses about underlying causal relationships between the dependent and independent variables.
- In Six Sigma projects most frequently used statistics is descriptive statistics and most useful distribution is normal distribution.

Descriptive statistics

- Data distribution
  Normal Distribution, other distributions
- Measures of central tendency
  Mean, Median, Mode
- Measures of dispersion
  Standard Deviation, Variance and Range

Normal Distribution

- Most data tends to follow the *normal distribution* or bell shaped curve. One of the key properties of the normal distribution is the *relationship* between the *shape of the curve and the standard deviation.*

- *99.73% of the area of the normal distribution* is contained between −3 sigma and +3 sigma from the mean. Another way of expressing this is that 0.27% of the data falls outside 3 standard deviations for the mean.

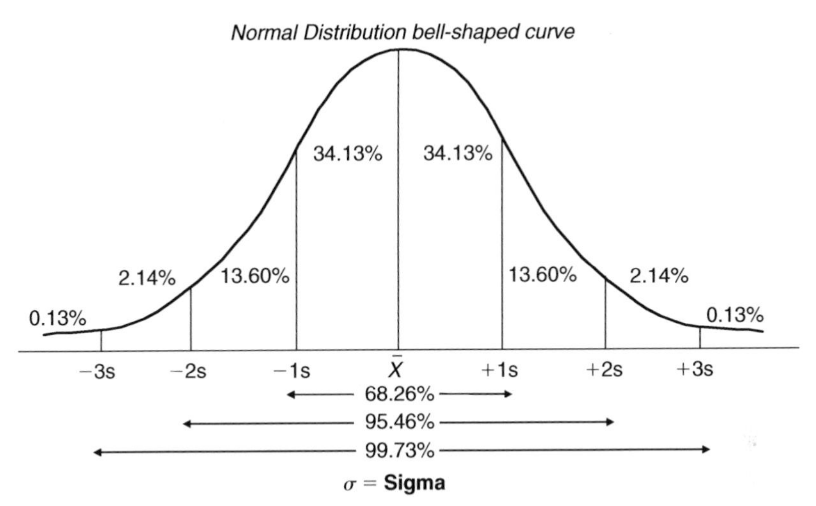

σ = **Sigma**

68.26% fall within +\−1 Sigma
95.46% fall within +\−2 Sigma
99.73% fall within +\−3 Sigma

$$\bar{X} = \frac{\sum_{i=1}^{n} X_i}{n-1}$$

Population is the total number in the study, e.g. census and a sample is a small subset of the population. Mean and Standard Deviation are expressed by Greek letters (μ, σ) in the population and by Roman letters (X, s).

Measures of central tendency

Measures of central tendency are determined by Mean, Median and Mode.

Mean is the arithmetic average of a set of data. It is a measure of central tendency, not a measure of variation. However, it is required to calculate some of the statistical measures of variation:

$$\text{Sample mean } (\bar{X}) = \frac{\sum_{i=1}^{n} X_i}{n-1}$$

where
$\bar{X}$ = Sample mean
X_i = Data point i
n = Group sample size

Median is the central or middle data point, e.g. in the series 32, 33, 34, 34, 35, 37, 37, 39, 41, the Median is 35.

Mode is the highest data point, e,g. in the series 32, 33, 34, 34, 35, 37, 37, 39, 41, the Mode is 41.

Measure of dispersion

The measure of dispersion is determined by Standard Deviation, Variance and Range and the shape of the distribution is measured by Skewness and Kurtosis.

Range is the simplest measure of dispersion. It defines the spread of the data and is the distance between the largest and the smallest values of a sample frequency distribution.

Variance is the average of the square values of the deviations from the mean.

Standard Deviation is defined as follows:

$$s = \sqrt{\sum_{i=1}^{n} \frac{(x_i - \overline{x})^2}{n-1}}$$

It is the square root of Variance and in a more intuitive definition, think of it as the 'average distance from each data point to the mean'.

Skewness and Kurtosis

Skewness measures the departure from a symmetrical distribution. When the tail of the distribution stretches to the left (smaller values) it is negatively skewed and when it stretches to the right it is positively skewed.

Kurtosis is a measure of a distribution peakness or flatness. A curve is too peaked when the kurtosis exceeds +3 and is too flat when kurtosis is less than −3.

Sources of variation

There are two types or sources of variation as shown in the following table:

Type	Definition	Characteristics
Common cause	No way to remove Influenced by one of the 6 Ms	*Always present* Expected Normal Random
Special cause	Can be removed Influence of the 6 Ms	*Not always present* Unexpected Not normal Not random

To distinguish between common cause and special causes of variation, use display tools that study variation over time such as Run Charts and Control Charts.

Process capability

- Process capability refers to the ability of a process to produce a product that meets given specification.
- It is measured by one or a combination of four indices:

DPM	Defects per million
σ level	Number of standard deviation from centre
C_p	Potential process index
C_{pk}	Process capability index

- σ level = minimum $(USL - \mu)/\sigma$, $(\mu - LSL)/\sigma$

  where USL = upper specification level
  LSL = lower specification level
  σ = standard deviation
  μ = mean
- C_{pk} = Process capability index
  = minimum $(USL - \mu)/3\sigma$, $(\mu - LSL)/3\sigma$
  = σ level/3

$$C_p = \frac{|USL - LSL|}{6\sigma}$$

Process Sigma

Process Sigma is an expression of process yield based on DPMO (defects per million opportunities) to define variation relative to customer specification. A higher value Process Sigma is an indication of a lower variation relative to specification.

Historical data indicates a change of Sigma value by 1.5 when long-term and short-term (actual) Process Sigma values are compared. The long-term values are higher by 1.5, e.g. 6 Sigma is long-term Process Sigma for 3.4 DPMO.

A sample Process Sigma calculation:

Given 7 defects (D) for 100 units (N) processed with 2 defect opportunities (O) per unit.

Defects per opportunity, DPO = $D/(N \times O) = 7/(100 \times 2) = 0.035$

Yield = $(1 - \text{DPO}) \times 100 = 96.5\%$

From the lookup table (see Appendix 2B);

- Long-term Process Sigma = 3.3
- Actual Process Sigma = 3.3 – 1.5 = 1.8

Inferential statistics

Inferential statistics helps us to make judgments about the population from a sample. A sample is a small subset of the total number in a population.

Appendix 2B
Yield Conversion Table

These are estimates of log-term sigma values. Subtract 1.5 from these values to obtain actual sigma values

Sigma	DPMO	Yield %	Sigma	DPMO	Yield %
6.0	3.4	99.99966	2.9	80757	91.9
5.9	5.4	99.99946	2.6	90801	90.3
5.8	8.5	99.99915	2.7	155070	88.5
5.7	13	99.99866	2.6	135666	86.4
5.6	21	99.9979	2.5	158655	84.1
5.5	32	99.9968	2.4	184060	81.6
5.4	48	99.9952	2.3	211855	78.8
5.3	72	99.9928	2.2	241964	75.8
5.2	108	99.9892	2.1	374253	72.6
5.2	159	99.984	2.0	308538	69.1
5.0	233	99.977	1.9	344578	65.5
4.9	337	99.966	1.8	382089	61.8
4.8	483	99.952	1.7	420740	57.9
4.7	687	99.931	1.6	460172	54.0
4.6	968	99.90	1.5	500000	50.0
4.5	1350	99.87	1.4	539828	46.0
4.4	1866	99.81	1.3	579260	42.1
4.3	2555	99.74	1.2	617911	38.2
4.2	3467	99.65	1.1	665422	34.5
4.1	661	99.53	1.0	691482	30.9
4.0	6210	99.38	0.9	725747	27.4
3.9	198	99.18	0.8	758036	24.2
3.8	10724	98.9	0.7	788145	21.2
3.7	13903	98.6	0.6	815940	18.4
3.6	17864	98.2	0.5	841345	15.9
3.5	22750	97.7	0.4	864334	13.6
3.4	28716	97.1	0.3	884930	11.5
3.3	35930	96.4	0.2	903199	9.7
3.2	44565	95.5	0.1	919243	8.1
3.1	54799	94.5			
3.0	66897	93.3			

Sample statistics are summary values of sample and are calculated using all the values of the sample. Population parameters are summary values of the population but they are not known. That is why we use sample statistics to infer population parameters.

Sample size (SS) = $(DC \times V/DP)^2$

where

DC = Degree of confidence = the number of standard errors for the degree of confidence

V = Variability = the standard deviation of the population

DP = Desired precision = the acceptable difference between the sample estimate and the population value.

A *null hypothesis* refers to a population parameter and not a sample statistic. Based on the sample data the researcher can accept the null hypothesis or accept the alternative hypothesis.

The *t-test* compares the actual difference between two means in relation to the variation in the data (expressed as the standard deviation of the difference between the means). It is applied when sample sizes are small enough (less than 30) that using an assumption of normality. A test of the null hypothesis that the means of two normally distributed populations are equal.

Analysis of variance (ANOVA) is used to assess the statistical differences between the means of two or more groups of population. ANOVA can also examine research problems that involve several independent variables.

The F-test assess the differences between the group means when we use ANOVA.

F = Variance between groups/variance within groups. Larger F-ratios indicate significant differences between the groups and most likely the null hypothesis will be rejected.

Causal modelling

Many business objectives are concerned with the relationship between two or more variables. Variables are linked together if they exhibit *co-variation*, i.e. when one variable consistently changes relative to other variable. *Co-efficient of correlation* is used to assess this linkage. Large co-efficients indicate high co-variation and a strong relationship and vice versa.

Causal modelling develops equations underlying causal relationships between the dependent and independent variables. Values of variables are plotted graphically and analysed by regression analysis and co-efficient of correlation to establish relationships between variables and their validity.

SPSS

The data analysis and statistical techniques presented in this appendix are all available in the popular software package SPSS (www.spss.com).

Appendix 3
Random nominal numbers

$\mu = 0, \sigma = 1$

	(1)	(2)	(3)	(4)	(5)	(6)	(7)
1	0.464	0.137	2.455	−0.323	−0.068	0.296	−0.288
2	0.060	−2.526	−0.531	−1.940	0.543	−1.558	0.187
3	1.486	−0.354	−0.634	0.697	0.926	1.375	0.785
4	1.022	−0.472	1.279	3.521	0.571	−1.851	0.194
5	1.394	−0.555	0.046	0.321	2.945	1.974	−0.258
6	0.906	−0.513	−0.525	0.595	0.881	−0.934	1.579
7	1.179	−1.055	0.007	0.769	0.971	0.712	1.090
8	−1.501	−0.488	−0.162	−0.136	1.033	0.203	0.448
9	−0.690	0.756	−1.618	−0.445	−0.511	−2.051	−0.457
10	1.372	0.225	0.378	0.761	0.181	−0.736	0.960
11	−0.482	1.677	−0.057	−1.229	−0.486	0.856	−0.491
12	−1.376	−0.150	1.356	−0.561	−0.256	0.212	0.219
13	−1.010	0.589	−0.918	1.598	0.065	0.415	−0.169
14	−0.005	−0.899	0.012	−0.725	1.147	−0.121	−0.096
15	1.393	−1.163	−0.911	1.231	−0.199	−0.246	1.239
16	−1.787	−0.261	1.237	1.046	−0.508	−1.630	−0.146
17	−0.105	−0.357	−1.384	0.360	−0.992	−0.116	−1.698
18	−1.339	1.827	−0.959	0.424	0.969	−1.141	−1.041
19	1.041	0.535	0.731	1.377	0.983	−1.330	1.620
20	0.279	−2.056	0.717	−0.873	−1.096	−1.396	1.047
21	−1.805	−2.008	−1.633	0.542	0.250	0.166	0.032
22	−1.186	1.180	1.114	0.882	1.265	−0.202	0.151
23	0.658	−1.141	1.151	−1,210	−0.927	0.425	0.290
24	−0.439	0.358	−1.939	0.891	−0.227	0.602	0.973
25	1.398	−0.230	0.385	−0.649	−0.577	0.237	−0.289
26	0.199	0.208	−1.083	−0.219	−0.291	1.221	1.119
27	0.159	0.272	−0.313	0.084	−2.828	−0.439	−0.792
28	2.273	0.606	0.606	−0.747	0.247	1.291	0.063
29	0.041	−0,307	0.121	0.790	−0.584	0.541	0.484
30	−1.132	−2.098	0.921	0.145	0.446	−2.661	1.045
31	0.768	0.079	−1.473	0.034	−2.127	0.665	0.084
32	0.375	−1.658	−0.851	0.234	−0.656	0.340	−0.086
33	−0.513	−0.344	0.210	−0.736	1.041	0.008	0.427
34	0.292	−0.521	1.266	−1.206	−0.899	0.110	−0.528
35	1.026	2.990	−0.574	−0.491	−1.114	1.297	−1.433
36	−1.334	1.278	−0.568	−0.109	−0.515	−0.566	2.923
37	−0.287	−0.144	−0.254	0.574	−0.451	−1.181	−1.190
38	0.161	−0.886	−0.921	−0.509	1.410	−0.518	0.192
39	−1.346	0.193	−1.202	0.394	−1.045	0.843	0.942
40	1.250	−0.199	−0.288	1.810	1.378	0.584	1.216

Appendix 4 Answers to numeric exercises

Part 2

6.

(b) Mean (μ) = $\Sigma y/20 = (21 + 19 + 20 + \ldots + 22)/20 = 400/20 = 20$

Standard Deviation

$$(\sigma) = \surd\left(\Sigma(y-\mu)^2/(n-1)\right) = \surd(32/19) = \surd(1.684) = 1.3$$

(c) $\text{UCL} = \mu + 3\sigma = 20 + 3 \times 1.3 = 23.9$

$\text{LCL} = \mu - 3\sigma = 16.1$

7. Nordanmatic is the control station with a maximum output rate of 150 cartons (with tubes) per minute

$$\begin{aligned}\text{Maximum speed} &= 150 \times 60 = 9000 \text{ units per hour}\\ \text{Maximum output in 8 hours} &= 8 \times 9000 = 72\,000 \text{ units}\\ \text{OEE} &= \text{Good output /Maximum output}\\ &= 39\,600/72\,000 = 0.55\\ &= 55\%\end{aligned}$$

10.

(a)

$$\text{Time at completion} = \frac{\text{Original Duration (OD)}}{\text{Schedule Performance Index (SPI)}}$$

SPI at week 18 = Earned value/Planned spend
= 130 000 / 140 000
= 0.93

Time at completion = OD/SPI = 50/0.93 = 53.8 weeks

$$\text{Estimated cost at completion} = \frac{\text{Budget at Completion (BAC)}}{\text{Cost Performance Index (CPI)}}$$

$$\begin{aligned} \text{CPI at week 18} &= \text{Earned value/Actual cost} \\ &= 130\,000\,/\,150\,000 \\ &= 0.867 \end{aligned}$$

Estimated cost at completion = BAC/CPI = £300 000/0.867 = £346 200

Part 3

3. Result
   95% Confidence Interval is 0.089 − 0.311.
   Sample size is 246.

References

Chapter 1

Basu, R. and Wright, J.N. (1997). *Total Manufacturing Solutions*. Butterworth Heinemann, Oxford.
Basu, R. and Wright, J.N. (2003). *Quality Beyond Six Sigma*. Butterworth Heinemann, Oxford.
Best Practices, LLC (2000). *Building Six Sigma Excellence, a Case Study of General Electric*. Oxford.
Dale, B. and McQuater, R. (1998). *Managing Business Improvement and Quality*. Blackwell Publishers, Oxford.
Easton, G. and Jarrell, S. (1998). The effects of total quality management on corporate performance. *Journal of Business*, 71, 253–307.
Foster, S.T. (2001). *Managing Quality*. Prentice Hall, Oxford.
Gravin, D. (1984). What does product quality really mean? *Sloan Management Review*, 25(2).
Hill, T. (2000). *Operations Management*. Basingstoke, UK, Palgrave.
International Who's Who of Professionals (2003). Vol 2, 8742, Jacksonville: WHO's WHO Historical Society.
Kano, N. (Ed.) (1996). *A Guide to TQM for Service Industries*. Asian Productivity Organisation.
Mamchak, P.S. and Mamchuck, R.S. (1999). *2002 Gems of Educational Wit and Humour*. Lincoln, UK, Parker Publishers.
Parasuraman, A., Zeithamel, V. and Berry, L. (1984). *A Conceptual Model of Service Quality*. Marketing Science Institute.
Parasuraman, A., Zeithamal, V.A. and Berry, L.L. (1985). A Conceptual Model of Service Quality and its implication for future research. *Journal of Marketing* 49, 41–50.
Taylor, F.W. (1929). *The Principles of Scientific Management*. New York, Harper & Bros.
Wild, R. (2002). *Operations Management*. Continuum, London

Chapter 2

Barney, M. and McCarty, T. (2004). *New Six Sigma*. Motorola University, London.
Basu, R. (2001). Six Sigma to Fit Sigma. *The IIE Solutions*. Atlanta. July 2003.
Basu, R. and Wright, J.N. (1998). *Total Manufacturing Solutions*. Oxford, Butterworth-Heinemann.
Basu, R. and Wright, J.N. (2003). *Quality Beyond Six Sigma*. Oxford, Butterworth Heinemann.
Carlzon, J. (1989). *Moments of Truth*. New York, Harper Row.
Crosby, P.B. (1979a). *Quality is Free*. New York, Mcgraw Hill.

Crosby, P.B. (1979b). *Quality without Tears*. In Turner, J.R. (Ed.) (1993). *The Handbook of Project Based Management: Improving the Processes for Achieving Strategic Objectives*. London, McGraw Hill.

Dale, B.G. (Ed.) (1999). *Managing Quality*. New York, Prentice Hall.

Davenport, T.H. (1993). *Process Innovation*. Boston, MA, Harvard Business School Press.

Davenport, T.H. and Short, J.E. (1990). The new industrial engineering: Information technology and business process re-design. *Sloan Management Review*, 31(4).

Deming, W.E. (1986). *Out of the Crisis*. In Wright, J.N. (Ed.) (1999). *The Management of Service Operations*. London, Cassell.

Feigenbaum, A.V. (1983). *Total Quality Control*. New York, McGraw-Hill.

Fry, T.D., Steele, D.C. and Sladin, B.A. (1994). A service orientated manufacturing strategy. *International Journal of Operations and Production Management*, 14(10), 17–29.

Gabor, A. (2000). He made America think about quality. *Fortune*, 142(10). October, New York

Hammer, M. and Champy, J. (1993). *Re-Engineering the Corporation*. London, Nicholas Brealey.

Harrington, H.J. (1987). *The Improvement Process*. New York, McGraw Hill.

Harrison, A. (1998). Manufacturing strategy and the concept of world class manufacturing. *International Journal of Operations and Production Management*, 18(4), 397–408.

Hayes, R.H. and Wheelwright, S.C. (1984). *Restoring our Competitive Edge: Competing through Manufacturing*. New York, John Wiley & Sons.

Imai, M. (1997). *Gemba Kaizen*. New York, McGraw Hill.

Ishikawa, K. (1979). *Guide to Quality Control*. Tokyo, Asian Productivity Organisation.

Ishikawa, K. (1985). *What is Total Quality Control? The Japanese Way*. Englewood Cliffs, Prentice Hall. Translated by Lu, D.J.

Juran, J.M. (1988). *Juran on Planning for Quality*. New York, Free Press.

Juran, J.M. (1989). *Juran on Leadership for Quality: An Executive Handbook*. New York, Free Press.

Knuckey, S., Leung-Wai, J. and Meskill, M. (1999). *Gearing Up: A study of Best Manufacturing Practice in New Zealand*. Wellington, Ministry of Commerce.

Marash, S.A. (2004). *Fusion Management*. Fairfax, Virginia, QSU Publishing Company.

Oakland, J.S. (2003). *Total Quality Management: Text with Cases*, 3rd Edition. Oxford, Butterworth Heinemann.

Plaster, G. and Alderman, J. (2006). *Beyond Six Sigma-Profitable Growth through Customer Value Creation*. Wiley, New York.

Pyzdek, T. (2000). The Six Sigma Revolution. www.pyzdek.com/six-sigma-revolution.htm (2 May 2002).

Schonberger, R. (1986). *World Class Manufacturing*. New York, Free Press.

Shingo, S. (1988). *A Revolution in Manufacturing: The SMED System*. Cambridge, MA, Productivity Press.

Shirose, K. (1992). *TPM for Workshop Leaders*. Cambridge, MA, Productivity Press.

Womack, J., Jones, D. and Roos, D. (1990). *The Machine that Change the World*. New York, Rawson and Associates.

Chapter 3

Basu, R. and Wright, J.N. (2003). *Quality Beyond Six Sigma*. Oxford, Butterworth Heinemann.

Dale, B.G., Booden, R.J. and Wilcox, M. (1993). *Difficulties Encountered in the Use of Quality Management Tools*. Manchester, UMIST.
McElroy, B. and Mills, C. (2000). *Managing Stakeholders, Gower Handbook of Project Management, Gower*. Aldershot, England, Gower Publishing. pp. 757–792.
McQuater, R.E., Dale, B.G., Wilcox, M. and Booden R.J. (1994). The effectiveness of quality management techniques and tools in the continuous improvement process, *Proceedings of Factory 2000*, Conference publication number 328, IEE, London, October, pp. 574–580.
Tuckman, B.W., Abry, D.A., Smith, D.R. and Arby, D. (2001). *Learning and Motivation Strategy*. New York, Prentice Hall.
Wallace, T.F. (1990). *MRPII: Making It Happen*. New York, John Wiley & Sons.

Chapter 4

Eckes, G. (2001). *The Six Sigma Revolution*. New York, John Wiley & Sons.
George, M.L. (2002). *Lean Six Sigma*. New York, McGraw Hill.

Chapter 5

Basu, R. and Wright, J.N. (2003). *Quality Beyond Six Sigma*. Oxford, Butterworth Heinemann.
Kolarik, W.J. (1995). *Creating Quality*. New York, McGraw Hill Inc.
Ledolter, J. and Burnill, C.W. (1999). John Wiley & Sons. Statistical QC, New York,
Schmidt, S.R., Kiemele, M.J. and Berdine, R.J. (1999). *Knowledge Based Management*. Colorado Springs, Academy Press.
Stamatis, D.H. (1997). *TQM Engineering Handbook*. New York, Marcel Dekker Inc.

Chapter 6

Altshuller, G. (1994). *And Suddenly the Inventor Appeared*, translated by Lev Shulyak, Technical Innovation Center. Worcester, MA.
Bassard, and Ritter (1994). *The Memory Jogger*. Massachusetts, Goal/QPC. p. 81.
Hartman, E. (1991). How to install TPM in your Plant, *8th International Maintenance Conference*, Dallas, 12–14 November.
Kotabe, M. and Helsen, K. (2000). *Global Marketing Management*. New York, John Wiley & Sons.
Moroney, M.J. (1973). *Fact from Figures*. Middlesex, England, Penguin Books.
Shirose, K. (1992). *TPM for Workshop Leaders*. Cambridge, Massachusetts, Productivity Press.
Stamatis, D.H. (1999). *TQM Engineering Handbook*. New York, Marcel Dekker Inc.
Turner, J.R. and Simister, S.J. (Eds) (2000). *Gower Handbook of Project Management*. Gower Publishing, Aldershot, England.
Wild, R. (2002). *Operations Management*. London, Continuum.

Chapter 7

Basu, R. and Wright, J.N. (1997). *Total Manufacturing Solutions*. Oxford, Butterworth Heinemann.
Buzan, T. (1995). *The Mind Map Book*. London, BBC Books.

Lewin, K. (1951). *Field Theory in Social Science: Selected Theoretical Papers*. New York, Harper Collins.
Schmidt, S.R., Kiemele, M.J. and Berdine, R.J. (1999). *Knowledge Based Management*. Colorado Springs, Air Academy & Associates.
Skinner, S. (June 2001). Mastering a basic tenet of lean manufacturing – Five S. *Manufacturing News, USA*, 8(11).
Womack, J.P. and Jones, D.T. (1996). *Lean Thinking*. London, Touchstone Books.

Chapter 8

Basu, R. and Wright, J.N. (1997). *Total Manufacturing Solutions*. Oxford, England, Butterworth Heinemann.
Basu, R. and Wright, J.N. (2003). *Quality Beyond Six Sigma*. Oxford, Butterworth Heinemann.
Juran, J.M. (1999). *Juran's Quality Handbook*, 5th Edition. New York, McGraw Hill.
Wild, R. (2002). *Operations Management*. London, Continuum.

Chapter 9

Choudhury, S. (2003). *Design for Six Sigma*. London, FT/Prentice Hall.
Churchman, C.W., Ackoff, R.L. and Arnoff, E.L. (1968). *Introduction to Operations Research*. New York, John Wiley & Sons.
Dale, B.G. (2000). *Managing Quality*. Oxford, Blackwell Publishers.
Hauser, J.R. and Clausing, D. (1988). *The House of Quality*. Harvard Business School Press.
Mizuno, S. and Akao, Y. (1994). *QFD: The Customer Driven Approach to Quality Planning and Deployment*. Cambridge, Massachusetts, Productivity Inc.
Saaty, T.L. (1959). *Mathematical Methods of Operations Research*. New York, McGraw Hill Company.
Waller, D.L. (2002). *Operations Management*. London, Thomson Learning. p. 157.

Chapter 10

Basu, R. (2001). *New Criteria of Performance Management*, Measuring Business Excellence 5.4. MCB University Press, Bradford.
Basu, R. and Wright, J.N. (2003). *Quality Beyond Six Sigma*. Oxford, Butterworth Heinemann.
Basu, R. and Wright, J.N. (2007). *Total Supply Chain Management*. Elsevier, Oxford.
Bromwich, M. and Bhimani, A. (1989). *Management Accounting: Research Studies*. London, CIMA.
Camp, R.C. (1989). *Benchmarking, the Search for Industry Best Practices that Lead to Superior Performance*. Milwaukee, ASQC Industry Press.
Colding, S. (1995). *Best Practice Benchmarking*. Basingstoke, Hants, Gower Press.
Dale, B.G. (1999). *Managing Quality*. Oxford, Blackwell Publishers.
The European Foundation of Quality Management (EFQM) Model. (1999). London: British Quality Foundations.
Goldratt, E.M. (1992). *The Goal*. North River Press, New York.
Johnson, T. and Kaplan, R. (1987). *Relevance Cost: The Rise and Fall of Management Accounting*. Boston, Harvard Business School Press.

Kaplan, R.S. (1996). *Implementing the Balanced Scorecard.* Harvard Business School Publishing.
Kaplan, R.S. and Norton, D.P. (1992). The Balanced Scorecard – Measures that drive performance. *Harvard Business Review.*
Kaplan, R.S. and Norton, D.P. (1996). *The Balanced Scorecard.* Harvard Business School Press.
Karlof, B. and Ostblom, S. (1994). *Benchmarking.* Chichester, UK, John Wiley & Sons.
Link, C.L. and Goddard, W.E. (1988). *Orchestrating Success.* New York, John Wiley & Sons.
Maskell, B.H. (1996). *Making Numbers Count.* Oregon, Productivity Press.
McLymont, R. and Zuckerman, A. (2001). Slipping into ISO 9000: 2000. *Quality Digest, USA*, August.
Ohno, T. (1988). *Toyota Production System.* Cambridge, USA, Productivity Press.
Sayle, A.J. (1991). *Meeting ISO 9000 in a TQM World.* UK, AJSL.
Shingo, S. (1988). *Non-Stock Production.* Cambridge, USA, Productivity Press.
Wallace, T.F. (1990). *MRPII: Making It Happen.* New York, John Wiley & Sons.
Welch, J. and Byrne, J.A. (2001). *Jack: Straight from the Gut.* Warner Books.
Womack, J.P. and Jones, D.T. (1996). *Lean Thinking.* London, Touchstone Books.

Chapter 11

Albrecht, K. (1988). *At America's Service.* Homewood, Il, Dow Jones-Irwin.
Axelrod, R.H. (2001). Changing The Way We Change Organizations. *The Journal of Quality and Participation*, 24(1), 22–27. Spring
Barlett, C. and Ghoshal, S. (1994). *The Changing Role of Top Management. INSEAD working papers.* London Business School Library.
Basu, R. and Wright, J.N. (1998). *Total Manufacturing Solutions.* Oxford, Butterworth-Heinemann.
Best Practice (1999). *Building Six Sigma Excellence: A Case Study of General Electric.* Chapel Hill, NC, Best Practice LLC.
Carnall, C. (1999). *Managing Change in Organisations.* London, Prentice Hall Europe.
CCTA (1999). *Central Computer and Telecommunication Agency. Managing Successful Programmes.* London, Stationery Office.
Collins, J.C. and Porras, J.I. (1991). Organizational vision and visionary organizations. *Californian Management Review*, 34, 30–52.
Creech, B. (1994). *The Five Pillars of TQM.* New York, Truman Talley Books.
Crosby, P.B. (1979). *Quality without Tears.* In Turner, J.R. (Ed.) (1993). *The Handbook of Project Based Management: Improving the Processes for Achieving Strategic Objectives.* London, McGraw Hill.
Dulewicz, V., MacMillan, K. and Herbert, P. (1995). Appraising and developing the effectiveness of Boards and their Directors. *Journal of General Management*, 20(3), 1–19.
El-Namki, M.S.S. (1992). Creating a corporate vision. *Long Range Planning*, 25(6), 25–29.
Foreman, S.K. (1996). *Internal Marketing.* In Turner, J.R., Grude, K.V. and Thurloway, L. (Eds). *The Project Manager as Change Agent.* London, McGraw Hill.
Foreman, S.K. (2000). *Internal Marketing.* In Lewis, B. and Varey, R. (Eds). *Internal Marketing.* Psychology Press, London.
Foreman, S.K. and Money, A.H. (1995). *Internal Marketing: Concepts, Measurement and Application.* Oxon, Henley Management College. Unpublished paper.

Friedman, M. (1970). A Friedman Doctrine: The social responsibility of business is to increase its profits. *New York Times Magazine*, September 13, 32.
Gabor, A. (2000). He made America think about quality. *Fortune*, October 142(10). New York.
Hall, J. (1999). Six principles for successful business change management. *Management Services*, 43(4), 16–18.
Hammer, M. and Champy, J. (1993). *Re-Engineering the Corporation*. London, Nicholas Brealey Publishing.
Ishikawa, K. (1985). *What is Total Quality Control? The Japanese Way*. Englewood Cliffs, Prentice Hall. Translated by Lu, D.J.
Knuckey, S., Leung-Wai, J. and Meskill, M. (1999). *Gearing Up: A study of Best Manufacturing Practice in New Zealand*. Wellington, Ministry of Commerce.
Kotler, P. (1999). *Kotler on Marketing*. New York, Free Press.
Langeler, G.H. (1992). The vision trap. *Harvard Business Review*, March–April, 46–49.
Machiavelli, N. (1513). *The Prince*. New York, New American Library of World Literature. Translated by Luigi, R., revised by Vincent, E.R.P. (1952).
Mintzberg, H. (1996). Musings on management. *Harvard Business Review*, July–August, 61–67.
Peters, T. and Austin, N. (1986). *A Passion for Excellence*. London, Fortune.
Peters, T. and Waterman, J.R. (1982). *In Search of Excellence*. New York, Harper Row.
Schonberger, R. (1986). *World Class Manufacturing*. New York, Free Press.
Stacey, R.D. (1993). *Strategic Management and Organizational Dynamics*. London, Pitman Publishing.
Turner, J.R., Grude, K.V. and Thurloway, L. (1996). *The Project Manager as Change Agent*. London, McGraw Hill.
Wright, J.N. (1996). Creating a quality culture. *International Journal of General Management*, 21(3), 19–29.
Wright, J.N. (1999). *The Management of Service Operations*. London, Cassell.
www.airacad.com (April 2002). Types of Culture: Six Sigma Training Notes. Colorado: Air Academy Associates.
www.minitab.com (May 2002).
Zeithaml, V.A., Parasuraman, A. and Berry, L.L. (1990). *Delivering Quality Service: Balancing Customer Perceptions and Expectations*. New York, The Free Press.

Chapter 12

Basu, R. and Wright, J.N. (2003). *Quality Beyond Six Sigma*. Oxford, Butterworth Heinemann.
Our Search for the Best (October 1992). *Industry Week*, USA.
Leading Manufacturing Practices (1994). Port Sunlight, UK: Unilever Research & Engineering Division.

Appendix 1

Ansoff, I. (1987). *Corporate Strategy*. London, Penguin Books.
Basu, R. and Wright, J.N. (1997). *Total Manufacturing Solutions*. Oxford, Butterworth Heinemann.
Belbin, R.M. (1985). *Management Teams: Why they Succeed or Fail*. London, Heinemann.
Mintzberg, H. (1990). *Mintzberg on Management: Inside our Strange World of Organisations*. New York, The Free Press.

Pascale, R.T. (1990). *Managing on the Edge: How Successful Companies Use Conflict to Stay Ahead.* New York, Simon and Schuster.
Porter, M.E. (1985). *Competitive Strategy.* New York, Free Press.
Senge, P. et al. (1994). *The Fifth Discipline Fieldbook: Strategies and Tools for Building a Learning Organisation.* New York, Currency.
Stern, J.M. and Shiely, J.S. (2001). *The EVA Challenge: Implementing Value Added Change in an Organisation.* New York, John Wiley & Sons.
Turner, J.R. and Cochrane, R.A. (1993). The goals and methods matrix. *International Journal of Project Management,* 11(2).
Wild, R. (2002). *Operations Management.* London, Continuum.

Glossary

ABC Analysis It is based on a Pareto Analysis grouping units usually according to the share of annual cost. Units having 80% annual cost are considered in the 'A' classification, units with the bottom 5% share are 'C' items and units with costs in between are in the 'B' category.

Activity Based Costing It is analysing the cost of an operation at each processing step. In addition to measuring direct costs, it covers bottlenecks, delays and other time related activities to highlight areas of inefficiencies in an operation.

Activity Network Diagram is a network analysis technique to allow a team to find the most efficient path and realistic schedule of a project by graphically showing the completion time and sequence of each task.

Affinity Diagram is used to generate a number of ideas by a team and then organise natural groupings among them to understand the essence of a problem.

Bar Chart also known as Gantt Chart, indicates scheduling activities. Horizontal bars show the various activities with the length of the bar proportional to the duration of a particular activity.

Benchmarking is rating an organisation's products, processes and performances with other organisations in the same or another business. The objective is to identify the gaps with competitors and the areas for improvement.

Best Practice Best practice refers to any organisation that performs as well or better than the competition in quality, timeliness, flexibility and innovation. Best practice should lead to world-class performance.

Black Belts are experts in Six Sigma methods and tools. Tools include statistical analysis. Black Belts are project leaders for Six Sigma initiatives, they also train other staff members in Six Sigma techniques.

BPR Business Process Re-engineering has been described as a manifesto for revolution. The approach is similar to taking a clean piece of paper and starting all over by identifying what is really needed to make the mission of the organisation happen.

Brainstorming A free wheeling group session for generating ideas. Typically a group meeting of about seven people will be presented with a problem. Each member will be encouraged to make suggestions without fear of criticism. One suggestion will lead to another. All suggestions, no matter how seemingly fanciful, are recorded and subsequently analysed. Brainstorming is useful for generating ideas for further detailed analysis.

Cause and Effect Diagram The Cause and Effect, fishbone or Ishikawa Diagram was developed by Kaoru Ishikawa. The premise is that generally when a problem occurs the effect is very obvious, and the temptation is to treat the effect. With the Ishikawa approach the causes of the effect are sought. Once the cause is known and eliminated, the effect will not be seen again. For example, working overtime is an effect, adding extra staff does not remove the cause. The question is what caused the situation that led to overtime being worked.

Check Sheet is a method of systematically recording data from historical sources or observations as they happen. The patterns and trends can be clearly detected and shown.

Continuous Improvement is always looking for ways to improve a process or a product, but not necessarily making radical step changes. If the basic idea is sound then building on it will improve quality. In Japan this is known as Kaizen.

Control Chart is a tool in Statistical Process Control to monitor the number of defects found in a product or a process over time and study the variation and its source.

COPQ Cost of Poor Quality. The cost of poor quality is made up of costs arising from: internal failures, external failures, appraisal, prevention and lost opportunity costs. In other words all the costs that arise from non-conformance to a standard. Chapter 11 discusses COPQ in some detail.

CRM Customer Relationship Management is the development of the database and strategies necessary to have the maximum client relationships in terms of quality, cost, reliability and responsiveness.

CTQs In Six Sigma CTQs are referred to Critical to Quality. This simply means the identification of factors that are critical for the achievement of a level of quality.

Cycle Time is the elapsed time between two successive operations or the time required to complete an operation.

DFSS Design for Six Sigma, see Chapter 9 for detailed discussion. The steps are Define, Measure, Analyse, Design and Validate.

DMAIC Is the cycle of Define, Measure, Analyse, Improve and Control, see Chapter 9 for detailed discussion.

DOE The process of examining options in the design of a product or service. Controlled changes of input factors are made and the resulting changes to outputs noted. Losses from poor design include not only direct loss to the company from reworking and scrap, but includes for the user downtime due to equipment failure poor performance and unreliability. Poor customer satisfaction will lead to further losses by the company as market share falls.

DPMO Defects per million opportunities. This is the basic measure of Six Sigma. It is the number of defects per unit divided by the number of opportunities for defects multiplied by 1 000 000. This number can be converted into a Sigma value. For example Six Sigma = 3.4 per million opportunities.

DRP Distribution Requirement Planning is the planning step in the supply chain to move finished goods from production or stock to the customer.

Earned Value Management or Earned Value Analysis is a project control tool for comparing the achieved value of work in progress against the project schedule and budget. It can be performed at the single activity level and by aggregating the results up through the hierarchy or work breakdown structure.

e-business Electronic-business is more than the transfer of information using information technology. e-business is the complex mix of processes, applications and organisational structures.

EFQM The European Foundation for Quality Management is derived from the American Malcom Baldridge Quality award. It is an award for organisations that achieve world-class performance as judged by independent auditors against a checklist. The checklist is detailed and extensive and covers: Leadership, People Management, Policy and Strategy, Partnerships and Resource, Processes, People Satisfaction, Customer Satisfaction, Impact on Society and Business Results.

ERP Enterprise Resource Planning is the extension of MRPII systems to the management of complete business functions including Finance and Human Resources.

Failure Mode and Effect Analysis FMEA was developed in the aerospace and defence industries. It is a systematic and analytical quality planning tool for identifying for new products or services, at the design stage, what could go wrong during manufactures, or when in use by the customer. It is an iterative process and questions asked/examined are:

1. What is the function
2. Potential failure modes
3. The effect of potential failure
4. Review of current controls
5. Determination of risk priority (occurrence, detection and severity of failure)
6. Identification of corrective actions to eliminate failures
7. Monitoring of corrective actions and counter measures.

First Pass Yield FPY, also known as RTY, is the ratio of the number of completely defects free without any kind of rework during the process units at the end of a process and the total number of units at the start of a process. The theoretical throughput rate is often regarded as the number of units at the start of the process. RTY/FPY is used as a key performance indicator to measure overall process effectiveness.

Fishbone Diagram see Cause and Effect Diagram.

FIT SIGMA™ also see TQM, Six Sigma and Lean Sigma. FIT SIGMA™ incorporates all the advantages and tools of TQM, Six Sigma and Lean Sigma. The aim is to get an organisation healthy (fit) by using appropriate tools for the size and nature of the business (fitness for purpose) and to sustain a level of fitness. FIT SIGMA™ is a holistic approach.

Flow Process Chart A flow process chart sets out the sequence of the flow of a product or a procedure by recording all the activities in a process. The chart can be used to identify steps in the process, value adding activities and non-value adding activities.

FMEA see Failure Mode and Effect Analysis.

Gantt Chart see Bar Chart.

Green Belts are staff trained to be Six Sigma project leaders, they work under the guidance of Black Belts, see Black Belts.

Histogram A histogram is a descriptive and easy to understand chart of the frequency of occurrences. It is a vertical bar chart with the height of each bar representing the frequency of an occurrence.

Input Process Output Diagram All operations or processes have inputs and outputs. The process is the conversion of inputs into outputs. Analysis of inputs should be made to determine factors that influence the process, e.g. input materials from suppliers meeting specification, delivery on time and so on.

Ishikawa see Cause and Effect Diagram.

ISO 9000 To gain ISO 9000 accreditation an organisation has to demonstrate to an accredited auditor that they have a well-documented standard and consistent process in place which achieves a defined level of quality or performance. ISO accreditation will give a customer confidence that the product or service provided will meet certain specified standards of performance and that the product or service will always be consistent with the documented standards.

JIT Just in Time was initially a manufacturing approach where materials are ordered to arrive just when required in the process, no output or buffer stocks are held, and the finished product is delivered direct to the customer. Lean Sigma incorporates the principals of JIT and now relates to the supply chain from supplier and supplier's supplier, through the process to the customer and the customer's customer.

Kaizen Kaizen is a Japanese word derived from a philosophy of gradual day-by-day betterment of life and spiritual enlightenment. This approach has been adopted in industry and means gradual and unending improvement in efficiency and/or customer satisfaction. The philosophy is doing little things better so as to achieve a long-term objective.

Kanban Kanban is the Japanese word for card. The basic kanban system is to use cards to trigger movements of materials between operations in production so that a customer order flows through the system. Computer systems eliminate the need for cards but the principle is the same. As a job flows through the factory, completion of one stage of production triggers the next so that there is no idle time, or queues, between operations. Any one job can be tracked to determine the stage of production. A 'kanban' is raised for each customer order. The kanban system enables production to be in batches of one.

KPIs Key Performance Indicators include measurement of performance such as asset utilisation, customer satisfaction, cycle time from order to delivery, inventory turnover, operations costs, productivity and financial results (return on assets and return on investment).

Lean Sigma Also see Just in Time (JIT). Lean was initially a manufacturing approach where materials are ordered to arrive just when required in the process, no output or buffer stocks are held, and the finished product is delivered direct to the customer. Lean Sigma incorporates the principals of Six Sigma, and is related to the supply chain from supplier and supplier's supplier, through the process to the customer and the customer's customer.

Milestone Tracker Diagram is used to show the projected milestone dates and the best estimated date on the week of the progress review on a single chart.

Mind Mapping is a learning tool for ordering and structuring the thinking process of an individual or team working on a focused theme. According to Buzan, the Mind Map 'harnesses the full range of cortical skills – word, image, number, logic, rhythm, colour and spatial awareness – in a single, uniquely powerful technique'.

Mistake Proofing Refers to making each step of production mistake free. This is also known as Poka Yoke. Poka Yoke was developed by Shingo (also see SMED), and has two main steps: (1) preventing the occurrence of a defect and (2) detecting the defect. The system is applied at three points in a process:

1. In the event of an error, prevent the start of a process
2. Prevent a non-conforming part from leaving a process
3. Prevent a non-conforming product from being passed to the next process.

Monte Carlo Technique is a simulation process. It uses random numbers as an approach to model the waiting times and queue lengths and also to examine the overall uncertainty in projects.

MRP (II) Manufacturing Resource Planning is an integrated computer based procedure for dealing with all of the planning and scheduling activities for manufacturing, and includes procedures for stock reorder, purchasing, inventory records, cost accounting and plant maintenance.

Mudas Muda is the Japanese word for waste or non-value adding. The seven activities that are considered are: Excess production, Waiting, Conveyance, Motion, Process, Inventory and Defects. For further details see Chapter 1.

Nominal Group Technique is a brainstorming tool for a team to come to a consensus of the relative importance of ideas by completing importance rankings into a team's final priorities.

Normal Distribution or Gaussian Distribution is a graph of the frequency of occurrence of a random variable. The distribution is continuous, symmetrical, bell shaped, and the two tails extend indefinitely.

OEE Overall Equipment Effectiveness is the real output of a machine. It is given by the ratio of the good output and the maximum output of the machine for the time it is planned to operate.

Pareto Wilfredo Pareto was a 19th century Italian economist who observed that 80% of the wealth was held by 20% of the population. The same phenomenon can often be found in quality problems. Juran (1988) refers to the vital few and the trivial many. The technique involves collecting data of defects, identifying which occurs the most and which results in the most cost or damage. Just because one defect occurs more often than others does not mean it is the costliest or should be corrected first.

PDCA The Plan, Do, Check, Act cycle was developed by Dr W.E. Deming. It refers to: Planning the change and setting standards, Doing – making the change happen, Checking that what is happening is what was intended (standards are being met), and Act, taking action to correct back to the standard.

Performance Charts or UCL/LCL. Upper control and lower control limits are used to show variations from specification. Within the control limits performance will be deemed to be acceptable. The aim should be over time to reduce the control limits. Thus control charts are used to monitor processes and the data gathered from the charts should be used to force never-ending improvements. These types of charts might also be known as Tolerance Charts.

Poka Yoke see Mistake Proofing.

Project A project is a unique item of work for which there is a financial budget and a defined schedule.

Project Management involves the planning, scheduling, budgeting and control of a project using an integrated team of workers and specialists.

Project Charter A Project Charter is a working document for defining the terms of reference of each Six Sigma project. The charter can make a successful project by specifying necessary resources and boundaries that will in turn ensure success.

Process Capability Process Capability is the statistically measured inherent reproducibility of the output (product) turned out by a process that are within the design specifications of the product.

Process Mapping Process Mapping is a tool to represent a process by a diagram containing a series of linked tasks or activities which produce an output.

PESTLE Political, Economic, Social, Technical, Legal and Environmental is an analytical tool for assessing the impact of external contexts on a project or a major operation and also the impact of a project on its external contexts. There are several possible contexts including:

- Political
- Economic
- Social
- Technical
- Environmental.

QFD Quality Function Deployment is a systematic approach of determining customer needs and designing the product or service so that it meets the customers needs first time and every time.

Qualitative Uses judgement and opinions to rate performance or quality. Qualitative attempts to 'measure' intangibles such as taste, appearance, friendly service, etc.

Quality Circles Quality Circles are teams of staff who are volunteers. The team selects issues or areas to investigate for improvement. To work properly teams have to be trained, first in how to work as a team (group dynamics) and secondly in problem solving techniques.

Quality Project Teams Quality Project Teams are a top-down approach to solving a quality problem. Management determines a problem area and selects a team to solve the problem. The advantage over a Quality Circle is that this as a focused approach, but the disadvantage might be that members are conscripted rather than volunteers.

Quantitative Means that which is tangible or which can be measured. For example, speedometer on a car measures and shows the speed.

Regression Analysis is a tool to establish the 'best fit' linear relationship between two variables. The knowledge provided by the Scatter Diagram is enhanced with the use of regression.

Rolled Throughput Yield RTY, see First Pass Yield (FPY).

RU/CS Resource Utilisation and Customer Service. Analysis is a simple tool to establish the relative importance of the key parameters of both Resource Utilisation and Customer Service and to identify their conflicts.

Run Chart is a graphical tool to study observed data for trends or patterns over a specific period of time.

5Ss These represent a set of Japanese words for excellent house keeping (Sein – Sort, Seiton – Set in place, Seiso – Shine, Seiketso – Standardise and Sitsuke – Sustain).

S&OP Sales and operations planning is derived from MRP and includes new product planning. Demand planning, supply review, to provide weekly and daily manufacturing schedules, and financial information. Also see MRP(II). S&OP is further explained in Chapter 10, see Figure 10.7.

Scatter Diagram These diagrams are used to examine the relationship between two variables. Changes are made to each and the results of changes are plotted on a graph to determine cause and effect.

Sigma Sigma is the sign used for standard deviation from the arithmetic mean. If a normal distribution curve exists one sigma represents one standard deviation either side of the mean and accounts for 68.27% of the population. This is more fully explained in Chapter 3.

SIPOC is a high level map of a process to view how a company goes about satisfying a particular customer requirement in the overall supply chain. SIPOC stands for supplier, input, process, output and customer.

Six Sigma Six Sigma is a quality system which in effect aims for zero defects. Six Sigma in statistical term means six deviations from the arithmetic mean. This equates to 99.99966% of the total population, or 3.4 defects per million opportunities.

SMED Single Minute Exchange of Dies. This was developed for the Japanese automobile industry by Shigeo Shingo in the 1980s and involves the reduction of changeover of production by intensive work study to determine in-process and out-process activities and then systematically improving the planning, tooling and operations of the changeover process. Shingo believed in looking for simple solutions rather relying on technology.

SoQ Signature of Quality is a self-assessment process supported by a checklist covering: Customer focus, Innovation, Personnel and Organisational Leadership, use of Technology, and Environment and Safety issues. It is useful in FIT SIGMA™ for establishing a company 'health' report.

SPC Statistical Process Control (SPC) uses statistical sampling to determine if the outputs of a stage or stages of a process are conforming to a standard. Upper and lower limits are set, and sampling is used to determine if the process is operating within the defined limits.

The Seven Wastes also see Mudas.

SWOT Strengths, Weaknesses, Opportunities and Threats is a tool for analysing an organisation's competitive position in relation to its competitors.

Tolerance Charts see Performance Charts.

TPM Total Productive Maintenance requires factory management to improve asset utilisation by the systematic study and elimination of major obstacles – known as the 'six big losses' – to efficiency. The 'six big losses' in manufacturing are breakdown, setup and adjustment, minor stoppages, reduced speed, quality defects and startup and shutdown.

TQM Total Quality Management is not a system, it is a philosophy embracing the total culture of an organisation. TQM goes far beyond conformance to a standard, it requires a culture where every member of the organisation believes that not a single day should go by without the organisation in some way improving its efficiency and/or improving customer satisfaction.

UCL/LCL see Performance Charts.

Value Analysis very often a practice in purchasing, is the evaluation of the expected performance of a product relative to its price.

Value Chain also known as Porter's Value Chain. According to Michael Porter the competitive advantage of a company can be assessed only by seeing the company as a total system. This 'total system' comprises both primary and secondary activities.

Value Stream Mapping VSM is a visual illustration of all activities required to bring a product through the main flow, from raw material to the stage of reaching the customer.

World Class World class is the term used to describe any organisation that is making rapid and continuous improvement in performance and who is considered to be using 'best practice' to achieve world-class standards.

Zero Defects Philip Crosby made this term popular in the late 1970s. The approach is right thing, right time, right place and every time. The assumption is that it is cheaper to do things right the first time.

Index